Astronomers' Universe

For further volumes:
http://www.springer.com/series/6960

Tony Buick

Orrery

A Story of Mechanical Solar Systems, Clocks, and English Nobility

 Springer

Tony Buick
Orpington
United Kingdom

ISSN 1614-659X
ISBN 978-1-4614-7042-7   ISBN 978-1-4614-7043-4 (eBook)
DOI 10.1007/978-1-4614-7043-4
Springer New York Heidelberg Dordrecht London

Library of Congress Control Number: 2013947408

© Springer Science+Business Media New York 2014
This work is subject to copyright. All rights are reserved by the Publisher, whether the whole or part of the material is concerned, specifically the rights of translation, reprinting, reuse of illustrations, recitation, broadcasting, reproduction on microfilms or in any other physical way, and transmission or information storage and retrieval, electronic adaptation, computer software, or by similar or dissimilar methodology now known or hereafter developed. Exempted from this legal reservation are brief excerpts in connection with reviews or scholarly analysis or material supplied specifically for the purpose of being entered and executed on a computer system, for exclusive use by the purchaser of the work. Duplication of this publication or parts thereof is permitted only under the provisions of the Copyright Law of the Publisher's location, in its current version, and permission for use must always be obtained from Springer. Permissions for use may be obtained through RightsLink at the Copyright Clearance Center. Violations are liable to prosecution under the respective Copyright Law.
The use of general descriptive names, registered names, trademarks, service marks, etc. in this publication does not imply, even in the absence of a specific statement, that such names are exempt from the relevant protective laws and regulations and therefore free for general use.
While the advice and information in this book are believed to be true and accurate at the date of publication, neither the authors nor the editors nor the publisher can accept any legal responsibility for any errors or omissions that may be made. The publisher makes no warranty, express or implied, with respect to the material contained herein.

Printed on acid-free paper

Springer is part of Springer Science+Business Media (www.springer.com)

To Chris, Jo, Tim, Kat, lovely Caitlin and Abigail who became one year old when creation of this work began.

Foreword

Charles Boyle, the 4th Earl of Orrery, who gave his name to the marvel of mechanical engineering that is the subject of this book, grew up at Knole, the house in Kent where I now live. Tenuous though this connection might seem, it is one of many links and allusions that, for me, make this book so extraordinarily rich.

In the early chapters, Tony Buick expertly surveys developments in astronomy, mathematics, and the making of scientific instruments, that paved the way for the orrery, a machine which demonstrates the relative movements of the planets. Subsequent chapters read more like a detective story, as Tony Buick traces the origin of the prototype 'orrery' to the partnership in the early 18th century of Thomas Tompion, 'the father of English watchmaking', and George Graham. It was Graham's interest in astronomy that inspired his geared model of part of the solar system. The Earl of Orrery then commissioned a London instrument-maker called John Rowley to make him a copy of Graham's device—and that is how, by a stroke of patronage, the machine acquired its name.

As the illustrations in this handsome book show, orreries are things of great beauty, as well as miracles of precision. Tony Buick's text is equally precise, making complicated scientific and mechanical concepts clear. And yet he never loses sight of the personalities who made it all possible, and the social, cultural and intellectual milieus in which they moved: to such an extent that you sense the sights, sounds and smells of life in 18th-century London, as well as the broad sweep of scientific history.

Knole
January 2013

Robert Sackville-West

Preface

Time and time again, the question pops up about the origin of the mechanical solar system models that demonstrate the relative motions and often the sizes of the planets. Who started all those models and made the leap from static displays to clockwork mechanisms? When I followed these questions I found that full and detailed answers were not easy to come by and led to thumbing through, virtually and physically, the archives of world history, no less, to extract and crystallize as much relevant knowledge as was available. Comparing sources often threw up contradictions and so a judgement had to be made as to which facts were the most likely. Internet public encyclopaedias have had bad press, but unfairly so. Comparison between them and world famous ones showed that they usually came out very well and even sometimes the best in terms of information and errors. There were certainly many real horrors, not just slips and misprints, in the classic volumes. Therefore, as many sources as possible had to be checked, national and international museums, auction houses, personal and family websites, churches, stately homes, old newspapers on line, Royal Society archives, ancient transactions of other societies, observatories, local councils (UK and other countries), London Guilds, Bank of England archives, libraries, commercial companies and much general browsing of the internet.

It is a fact that a mechanical, clockwork model of a small part of the solar system was first made around the early part of the eighteenth century, and it was made by George Graham. But did he make it on his own? Where did he get the money to do it? Was he commissioned? Who made the second one and the third, and where did they go before, fortunately, ending their days in the good hands of museums and other such care homes? Why did they become known as orreries?

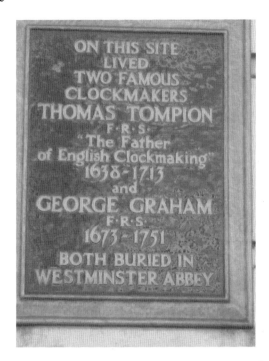

Fig. 1 Blue Plaque commemorating the dwelling place of George Graham and Thomas Tompion. Photograph by the author

The name of Eugene of Savoy kept cropping up, but what would an Austrian Prince have to do with anything, and how was the Duke of Marlborough involved, or his battles, a monastery and Irish nobles? So many questions! But there are answers, and the story had to start with scientific instruments and move on to clocks before focusing on the model itself. Surrounding the story are insights into the lives of revered scientists, Astronomers Royal, war leaders, monarchs and nobility, all of whom influenced the birth and life of the orrery.

The year 2013 marks the tercentenary of the death of Thomas Tompion, the master clockmaker who is such a crucial part of the story of the orrery. It is also the 340th anniversary of the birth of George Graham, 'Mr Orrery (Tellurion)' himself, and 320th of the birth of John Harrison, 'Mr Longitude'. All three were significant scientists of the age of enlightenment who joined with many others to make great contributions to the progress of science. A Blue Plaque commemorates Tompion and Graham at the corner of Fleet Street and Whitefriars (previously Water Lane) in London (Fig. 1). This year is also the 350th anniversary of the remarkable

Prince Eugene of Savoy, who was linked with the commissioning of the second proto-orrery; he collected fine treasures and fought alongside the Duke of Marlborough.

The beginning of the age of enlightenment is credited to Benedict de Spinoza (1632–1677), a Jewish-Dutch philosopher who sacrificed much in life to develop his thoughts about reason, such as refusing rewards, honors and prestigious appointments and giving his family inheritance to his sister. To him, everything must be looked at from an impartial perspective without coercion from religious scriptures or other preconceived ideas. Because of this approach, the age is also known as the age of reason. Spinoza, in addition to being a business man, was also a master grinder of lenses of such quality that Constantinjn Huygens (1628–1697; Christiaan Huygens' older brother) ground a 'clear and bright' 42 ft telescope lens from one of Spinoza's grinding dishes 10 years after his death. John Locke (1632–1704) and Pierre Bayle (1647–1706) were also early influences for the progression of enlightenment, or reason, followed by Newton and Voltaire. These men developed their ideas often at considerable risk to themselves.

To lay out the history of the orrery and mechanical clocks beginning solely with the seventeenth century, or maybe two centuries earlier, would be to ignore the elephant in the room. This would be the Antikythera mechanism. The mechanism's precision and wealth of understanding boggles the mind, not just the complexity of wheels, teeth and cogs, but the whole mathematics of the known Solar System. Questions remain. Why has only this one piece of rock-embedded, sophisticated mechanism been found? Will many more pop up out of chance discoveries and fill in two thousand years of hitherto unknown advanced technology? The phrase 'it will rewrite history' is often used to describe minor discoveries, but the uncovering of a plethora of Antikytheras and their successors really would put the cat amongst the pigeons! Nevertheless, there is only the 'modern' orrery and clock story of invention and development to be told by bringing together morsels of written anecdotes, records of provenance and delving into the history of, mostly but not entirely, the UK, Europe and the USA.

Rarely does one invention pop out of the blue with no previous gradual progression and the orrery is no exception. Moving machines have fascinated scientists from time immemorial, albeit with a small number of parts (except possibly the Antikythera

FIG. 2 A fine red umbra of an eclipse of the Moon. Photograph by the author

mechanism), and hence the story must begin with primitive instruments such as shadow sticks, sundials and water-driven clocks, from where it moves on to instruments requiring the precision engineer to meet ever-increasing demands. Einstein is quoted as having said, "Imagination is more important than knowledge. Knowledge is limited; imagination encircles the world." Also, "When I examine myself and my methods of thought, I come close to the conclusion that the gift of fantasy [imagination] has meant more to me than my talent for absorbing absolute knowledge." And so it must have been that, with George Graham's imagination firmly spread throughout the heavens above, it eventually came back down to Earth with a design for an educational model to indicate the regular and relative motions of the Sun, Earth and Moon. Not least amongst the demonstrations available from an orrery is the eclipse of our Moon and the time of its arrival. It is a beautiful sight, as illustrated in Fig. 2. The eclipsed Moon can be very dark, almost invisible, or a stunning red color, as for the image captured in March 2007.

During the eighteenth century many scientists and astronomers recognized the attraction of producing orreries, not just as an educational tool, but to show off their talents and increase

Fig. 3 A modern view of Fleet Street. Photograph by the author

their business profile. Many instruments were produced with extraordinary complexity and precision, and many had extravagant adornments to enhance their appeal.

A gizmo needs a catchy label to popularize and adopting the title of a revered nobleman for the name was brilliant! Possibly the most common question when introducing the model is "who was Orrery?" The answer reveals the amazing histories of the Earls of Cork and Orrery and their influence on science, battles, politics and philosophy. Although the family has its roots in Ireland, it spread throughout Britain with some concentration in the southern counties of England. They became so involved in top class academic squabbles that Charles Boyle ended up figuring prominently in the Phalaris question.

London was an acknowledged center of excellence for clockmakers in the seventeenth and eighteenth century, and the epicenter of that excellence was Fleet Street. A modern view is shown in Fig. 3. Not only did some of the best clockmakers live and work there, but famous scientists constantly visited the artificers and coffee houses to discuss science and business. Notable among them were giants such as Newton and Hooke, from whom much history is gleaned through his diaries. While the excellence of precision clock mechanics was honed in Fleet Street, the other main site of events was Greenwich, where a concentration of politics and practical application of astronomy was being played out.

The skills of the clockmakers were in great demand to serve the aspirations of those who wished to accurately study the stars or pursue their fortunes, recognition and, just maybe, the solution to the longitude problem. Even here the greatest scientists squabbled sometimes, resulting in slowing down the progress of science.

It is hard to imagine what life was like for the inhabitants of Fleet Street (and the rest of the ordinary folk of the country), not just scientifically, but every-day life with the squalor of the streets, crime and the lack of medicine that was accepted as the norm. Many would have witnessed major natural events such as storms and aurorae that could not be accounted for at the time. Scientists did their best to explain anything that occurred, even proposing a hollow Earth with inhabitants! Ridiculous now, but how can you have a hypothesis come true if you don't have any hypotheses to begin with!

Although the orrery's story is a great story to tell simply by itself, some of the technicalities are addressed for those who wish to delve a little deeper, such as meshing cogs, the equation of time and an explanation for the relative positions of the Solar System bodies.

Above all, the travails of the first orreries themselves are the centerpiece around which the other histories hang.

One cannot hope to cover every detail surrounding every relevant fact, especially as historical and scientific archives, events and artifacts are constantly being revealed or discovered. Hopefully, this work will stand as an overview for some and an introduction for those who wish to know more.

A glossary is included at the end of the work, but does not include words ubiquitously used throughout such as orrery and tellurium. Likewise for the index, where indications of the location of a word's appearance will be obvious from the contents.

Orpington, Kent, England   Tony Buick fecit
2012

About the Author

Dr. Tony Buick, FRSC, CChem, is an analytical chemist by profession specialising in several aspects of chromatography, High Performance Liquid Chromatography in particular, with international pharmaceutical, veterinary and agricultural companies. He is the author of the first and second editions of "How to Photograph the Moon and Planets with your Digital Camera" (Springer, 2006 and 2011) and "The rainbow Sky" (Springer 2010) that reflected his career interest in many aspects of spectroscopy including magnetic and electron spin resonance. He has had articles published in astronomy and various other magazines such as The Sky at Night and the Society for Popular Astronomy with regard to capturing photographs of the ISS and Iridium satellite trails, Transient Lunar Phenomena and lunar landscapes. He entered retirement by realising his ambition of teaching and has encouraged young and old to observe and understand the sky, especially while teaching science, computing and geography at a local school. Indeed, it was at that school where he showed the children at his science club how to make a human orrery and demonstrated the construction of an orrery from bits and pieces found around the house. His fascination with the orrery led to the research that forms the foundation of this book. His passion covers not only the infinite, looking through a telescope, but also the infinitesimal, looking through a microscope, and has published articles on tardigrades, robust microscopic animals that can even survive in space. The photography of wildlife is a recent pursuit with the publication of articles and photographs in wildlife magazines.

Acknowledgements

I am most fortunate to have a great friend, who knows just about everything astronomical, to read through the final manuscript. Gilbert Satterthwaite had many associations with the Royal Observatory, Greenwich, not least having been employed there under the tenth Astronomer Royal, Sir Harold Spencer Jones, during the 1950s. He was for many years very active in the history of astronomy, including seven years as Chairman of the Society for the History of Astronomy (later Vice Chairman). I am most grateful for his generous comments and the huge amount of his valuable time spent in indicating improvements and making some corrections. Any errors that may remain are entirely my responsibility. As 'Orrery' was about to go to print I received the sad news that Gilbert Satterthwaite had died. My gratitude for his support is deep. Gilbert was a true expert and friend and generous with giving his valuable time to astronomers of every level. He read and thoroughly checked every one of my astronomy books including 'Orrery' for which he gave a glowing report—"This is a book that needed to be written". I, and others around the world, will miss him greatly.

I am grateful and honoured that Robert Sackville-West, an ancestor of those who cared for the young Charles Boyle, the 4th Earl of Orrery, after whom was named the mechanical model, has been kind enough to provide the foreword. Robert and his ancestors have resided in and maintained the stunning Knole House that is mentioned and illustrated in the following chapters.

I am grateful to Dr Stephen Johnston of the Museum of the History of Science, University of Oxford, England, for leads into some vital proto-orrery references.

While many images have been acquired for use at some expense a few critical ones were freely provided or at reduced rates and I am grateful to the Chicago Adler Planetarium, the Oxford

Museum of the History of Science, the Armagh Planetarium in Northern Ireland, Knole House, the National Maritime Museum (Royal Observatory) and the British Museum. Many have allowed me to reproduce my own photographs of their treasures including the London Science Museum, the Collegiate Church in Youghal, County Cork, Ireland, and St. Paulinus church, Crayford, England.

Without the encouragement of Springer's John Watson this work would not have been written. Many thanks go to John for always being available for advice and consultation and Maury Solomon and Nora Rawn who patiently and professionally supported me to the final stages of the manuscript.

Support and patience from family and friends has been crucial. Tim, Kat, Chris, Jo, Caitlin and Abigail have been wonderful. And finally, Eileen has held cameras, steadied tripods, toured 'interesting' places with me, scoured books in libraries, spotted great subjects for photographs, given down-to-earth advice and criticism and generally supported all of the work from start to finish. Thank you Eileen!

Sir Patrick Moore CBE, FRS, FRAS; a Personal Memory

It is with much sadness that, as this book was being prepared, Sir Patrick Moore died—on the 9th December 2012. He must have had a million loyal and grateful friends and I feel honoured to be amongst that million. We were with him, Eileen and I, in his garden just a few weeks before when he was making sure we would be attending his usual New Years Eve party. Each such gathering, party or one of the many celebrations of his achievements, was marked by learning from so many how Patrick had changed or rescued their lives. They were not just astronomers but all sorts of people, scientists and non-scientists, who had been treated badly by life and were grateful to Patrick for his help. He never asked for thanks but he really appreciated loyalty. His generosity was immense and it was common for him to publish a book with ALL proceeds going to charities or local services.

At one of my visits an amateur filming crew asked Patrick to read from a card that took about 5 min. Unbeknown to all (although I was aware) his lovely black cat Ptolemy pawed at a cable resulting in an absence of sound on the final video. Patrick agreed to repeat but rejected the card since he had memorised it at one reading and seamlessly repeated the clip with the same enthusiasm. Such a good example of his intellectual brilliance! Oftentimes, while at his house, I would witness him giving impromptu and eloquent interviews over the phone. He always knew what to say and had the facts at his finger tips. One of his favourite moments to relate was when he played the piano accompaniment for Albert Einstein on his violin. He loved telling this story ending with "How I wish I had a video of that".

To some who didn't know him he gave the impression of being curt and xenophobic. Although he was fiercely patriotic I con-

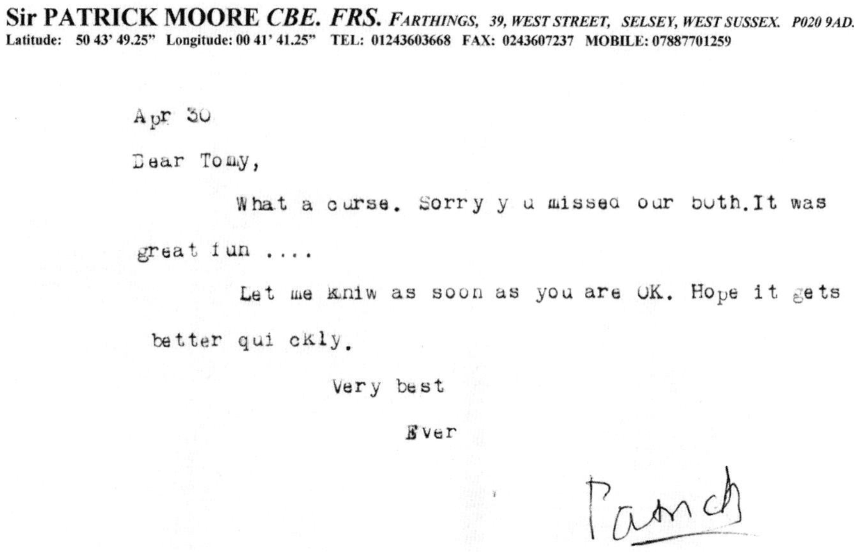

Sir Patrick Moore CBE, FRS, FRAS; a Personal Memory

Sir PATRICK MOORE CBE. FRS. FARTHINGS, 39, WEST STREET, SELSEY, WEST SUSSEX. PO20 9AD.
Latitude: 50 43' 49.25" Longitude: 00 41' 41.25" TEL: 01243603668 FAX: 0243607237 MOBILE: 07887701259

```
Apr 30
Dear Tony,
         What a curse. Sorry you missed our both. It was
great fun ....
         Let me kniw as soon as you are OK. Hope it gets
better quickly.
                   Very best
                        Ever
```

Patrick

FIG. 4 A note from Patrick

stantly saw him being hospitable, kind and helpful to so many from every part of the world.

I got to know him when I cheekily wrote, over a decade ago, to say there were things that he missed in his programme, The Sky at Night. He responded by inviting me to his home to discuss my photographs of the Moon. I have a pile of letters he wrote to me over the years but the original ones tapped out on his Woodstock typewriter were favourite. The note in Fig. 4 was typical, when I couldn't attend one of his celebrations due to a sore throat. He showed his concern, as always.

I was honoured when he used my photographs of his two cats modelling his antique orrery for his book 'Miaow', published this year, of which he was very proud. My copy was signed by him at our visit, Fig. 5, just a few weeks before he died.

If Patrick's belief in a life hereafter is correct he will now be enjoying paradise with his fiancée, Lorna, who has been waiting for him since they were young.

Thank you Patrick for support for all my books and being such a friend to a minion of the astronomical world!

Ever,
Tony

Fig. 5 At Patrick's home in late September, 2012, sharing orrery discussions with Ptolemy

Contents

1 Setting the Scene 1

2 Honest George, Chronometers and the Mystery of the Disappearing Proto-Orreries 53

3 Orrery—the Man and the Model 97

4 A Closer Look at Gear Calculations, Time Corrections, Escapements, and Orbital Resonance 149

5. The Clockmaker's London 199

6 Modern and Orrery Times Compared 253

Appendix 1: A Select Timeline 289

Appendix 2: Glossary 293

Appendix 3: Bibliography 295

Index 297

1. Setting the Scene

What Is an Orrery?

An orrery is a working model that demonstrates the movement of the planets, often with their moons, within our Solar System. More details of the eighteenth-century example shown in Fig 1.1 are given later.

Although the principle of making one or two spheres revolve around another seems simple by modern-day standards, the first ones were works of technical and imaginative brilliance at a time of striving for ever better micro-engineering precision. Much of the effort during the early-to mid-eighteenth century was driven by the imperative to produce the most accurate clocks for seafarers and their navigation, and some clockmakers became the orrery makers. But the first such devices were not called orreries! They were tellurions, telluriums, or planetaria. They were later named after the fourth Earl of Orrery.

So how did nobility become involved for the tellurion to be forever labeled orrery? The names of key figures of the time, Graham, Rowley and Tompion, do not fall off the tongue lightly for most. That is even true for those familiar with 'orrery,' but theirs is the credit for the essential dedication and skills developed to produce the attractive and appealing instruments. On its own the model was a small development, even a sideline, in the quest for better craftsmanship, accuracy and plain showing-off. But it sits at a key point within the story of astronomy from early humans staring at the sky in wonder and curiosity with an urge to make sense of all they saw right up to the modern-day achievements of space technology and understanding.

From scientific and lay observations there is no doubt about the major structure of the Solar System; local space travel and telescopes allow remote viewing of Earth and other planets to confirm its characteristics. But there was a time when such facts were not known, and hypotheses, religious pronouncements,

FIG. 1.1 Example of an eighteenth-century orrery. (Photograph by the author with permission of Sir Patrick Moore.)

myths and even guesses had to be made to account for the appearance and movement of the ever-present and constantly winking dots in the sky above. It is possible that the very first time anything 'up' in the night sky was noticed was when our hominid species advanced to the more energy-efficient walking on two legs (bipedalism) instead of four (quadrapedalism), to take advantage of the vast savannah lands that replaced forests as climate changes took place. That would be around four million years ago, when the Australopithecus genus roamed and evolved in East Africa before spreading far and wide and then becoming extinct, but not before its branches played a significant role in the evolution of modern man, Homo sapiens sapiens.

Astronomy in Ancient Civilizations

Human curiosity and understanding had civilizations throughout the world heading in the same direction in terms of wanting to account for the movement of the stars, planets, Sun and Moon and knowing what it meant for their religions, destiny, surviving nature's regular events and, naturally, personal power. To achieve that understanding, structures were built, simple and complex, to assist with observations and predictions. Outstanding among the ancient monuments is the World Heritage Site of Stonehenge in

FIG. 1.2 World Heritage Site of Stonehenge in Wiltshire, England. (Credit: Image from the Atlas van Loon of 1649, Public domain.)

Wiltshire, England, which was possibly built by the people of the time of 3100 B.C. to define the summer and/or winter solstices through the alignment of particular stones with the Sun and the Moon. Possibly it was an astronomical observatory or maybe a religious monument; it might even have been a place of healing or a burial site. Recent studies suggest the possibility that the orientation of the stones was more to do with access to a river than alignment with the Sun. Figure 1.2 is from the *Atlas van Loon 1649*. But even earlier than this, celestial objects proved to be fascinating for the very first high cultures we know of.

The wonderful and ancient names of Mesopotamia and Sumer began to emerge as archaeological evidence was unearthed dating from 5,000 B.C. onward to reveal the creation of a culture, very different from our own, building cities and forming a thinking society commonly described as the Cradle of Civilization. Mesopotamia, between the rivers, is a toponym for the Tigris-Euphrates river system corresponding to modern-day Iraq.

The earliest language used in Mesopotamia was Sumerian, and the early writing was cuneiform, or wedge-shaped, script. In fact, as early as 8000 B.C., with the development of pictograms (picture writing), records were kept—clay tokens—that eventually led to the cuneiform writing. Three wedges then a drawing

4   Orrery

Fig. 1.3 Sumerian inscription on an ancient cuneiform tablet. (Courtesy of the Schoyen Collection, public domain.)

of a bird meant three birds. Ten was represented by a circle. It was found by the Sumerians that wet clay could be neatly and accurately imprinted to replace scratching on stones, and since the cut reeds used as styluses best produced the shape of a triangle, or wedge, this became a basis of writing.

Cuneiform texts and artifacts that date back to about 6000 B.C. have been found in the valley of the Euphrates and are the earliest known attempts to catalog the stars and show star groupings such as the Lion and the Bull.

Baked clay was hard and very durable, hence the huge number of artifacts that have been unearthed and contributed to the collection of important archaeological evidence. Figure 1.3 shows a cuneiform tablet and the text is a list of gifts from the High and Mighty of Adab (an ancient city located in modern-day Iraq) to the High Priestess on the occasion of her election to the temple. It has clearly been written by an expert scribe.

The Sumerians and Babylonians developed great skills at mathematics, astronomy and, essential for that time, astrology.

Mathematics and science were based on a sexagesimal numerical system, i.e., to the base of 60. They multiplied 60 by 10, then multiplied 600 by 6, and so on. The number 60 has the advantage of being divisible by 2, 3, 4, 5, 6, 10, 12, 15, 20, and 30. The Sumerians also divided the circle into 360°. From these early people came the word "dozen" (a fifth of 60) and the division of the clock to measure hours, minutes, and seconds. Hence the 60-minute hour, 24-hour day and the 360° circle.

Since the survival of society depended so much on the seasons and natural events, it wasn't long before scientific skills were used to predict weather cycles, river behavior and any other regular feature that affected the cultivation of crops. Of course, the Egyptians were developing scientific and mathematical awareness at the same time, from around 5000 B.C. For them it was vital to predict the annual flooding of the Nile. Both the Babylonians and the Egyptians worked with the 365-day year.

It is not unexpected that both Mesopotamians and Egyptians would look to the night sky to predict events. Stars were grouped into patterns, and their orderly appearance clearly tied in with their observations of the natural cycles of events. Models, structures, stone circles and drawings were all created to assist in the correct prediction of events and times for festivals. In many cases the astronomer's/astrologer's reputation and life depended on the accuracy of these predictions.

The ancient Egyptians made many great advances in science, especially in medicine and alchemy. The Egyptians also contributed to ancient astronomy and, as with the Mesopotamians, their work was based upon agriculture and predicting the seasons. Since their very existence depended on it, the annual flooding of the Nile was the foundation of Egyptian civilization and agriculture, so predicting this occurrence with accuracy and in relation to their religion was the driving force behind the development of Egyptian astronomy.

The way the Egyptians built their magnificent pyramids (in a sense, their orreries or planetaria) and accurately aligned them with stars they were familiar with is a testament to the relationship they had with the sky and its mysteries. They seemed to be obsessed with true alignment with north/south, and a plethora of studies over many years have suggested how it was achieved—

from the position of the Sun and stars. Of course, it is important for modern studies to make allowances for the change in the relative positions of the stars as viewed from Earth and the various precessions of Earth since Egyptian times. The mathematical methods of the time, such as multiplication and division, involved basic empirical processes that were used by mathematicians without really knowing why they worked.

The history of science and technology in the Indian subcontinent begins with prehistoric human activity at Mehrgarh (one of the most important Neolithic, 7000–2500 B.C., sites in archaeology). Mehrgarh is in present-day Pakistan and continues through the Indus Valley. The oldest extant text of astronomy is the treatise by the Indian astronomer Lagadha, possibly dating to the last millennium B.C. or a little earlier. It describes rules for tracking the motions of the Sun and Moon. However, it was soon supplanted by the knowledge of the Greeks.

The ancient history of India is much about the Indus Valley civilization, 2500–1800 B.C., in the northwest. Many meticulously planned and constructed cities were found around the valley of the Indus River, where its people, the Harappans, evolved some new techniques in metallurgy and worked with copper, bronze, lead and tin. The engineering skill of the Harappans was remarkable, especially in building docks after a careful study of tides, waves and currents. They had their own script, the Indus Valley script, which is, as yet, undeciphered and may eventually reveal more of their secrets and science.

By the sixth century A.D. Indian astronomy and mathematics had become quite sophisticated. One of the main contributors was Aryabhata, a mathematician and astronomer who introduced the decimal point, arithmetic and geometric progressions, a method for determining the positions of the planets and the rotation of Earth on its axis. He possibly even suggested heliocentricism (where Earth and the other planets revolve around the Sun, not everything around Earth). He gave a value of π as

100+4, ×8, add 62,000, divide by 20,000=3.1416.

This was a pretty accurate value. A statue of Aryabhata is shown in Fig 1.4, although we don't know how true to life it is as there is no known information regarding his appearance.

FIG. 1.4 Aryabhata, an Indian mathematician and astronomer of the sixth century A.D. (Creative Commons, author unknown.)

It appears that even isolated early civilizations were developing their astronomical observation skills at about the same time as the brain was evolving, growing larger, and becoming more complex. The Chinese believed that the objects and happenings in the sky were linked to their destiny and almost every dynasty from the sixteenth century B.C. to the nineteenth century A.D. retained its official astronomer to observe and record changes in the heavens. Because of this there is a huge legacy of observational facts, most of which have been verified as accurate.

One early and tantalizing observation was that the Sun sometimes became much dimmer, and there was concern that the brightness might not return. So, meticulous records of time and size of the shadow were maintained. The earliest solar eclipse record that has been verified appears in a bone inscription dating back to the Shang dynasty, which ruled in the Yellow River valley in the second millennium B.C. Studies have proved that the solar eclipse recorded there actually took place on May 26, 1217 B.C., thus also proving that it was the first reliable record of an eclipse people ever made.

Records of lunar eclipses, however, date back to an even earlier time. Bone and tortoise shell inscriptions record five lunar eclipses that took place during the fourteenth and thirteenth centuries B.C. It is possible that the earliest records of sunspots were made by the Chinese in 28 B.C. Even a record of a solar prominence has been found inscribed on tortoise shell. Other observations within the records include novae, supernovae and comets, notably one that later became confirmed as Halley's Comet.

In spite of all the early Asian awareness of astronomy it is surprising that they maintained their belief in a flat Earth until introduced, by westerners in the seventeenth century, to the concept of a round one. This is a good example of the early isolation of civilizations and eventual exchange of information (theft?) and ideas.

It is amazing that so much brilliant science, mathematics and astronomy was spawned in the early civilization of Greece in spite of constant wars, internal and external. The highly organized culture of the Mycenaeans existed until around 1200, when it disappeared either by natural catastrophe or conquest possibly by the Dorians (Dorian is a term invented by historians to refer to that time).

This period was followed by the Dark Ages, 1200–800 B.C., so called partly because of the dearth of information and evidence to describe the period. There was huge economic growth from 800 B.C., resulting in an explosion of the size of families. Fighting continued between factions or from outside until around 510 B.C., the start of democracy, when it was decreed that all citizens should share in the political power. However, battles continued: the Battle of Marathon (490 B.C.); the Greco-Persian wars that continued until 449 B.C.; the Peloponnesian War, 431–404 B.C.; the Corinthian War, 395–387 B.C.; and so on for centuries. In spite of this, scientific and astronomical intellectual giants flourished such as Pythagoras, Archimedes and Hipparchus, who made enormous contributions to the development of understanding of the natural world.

On the American continent the Mayan civilization ranked highly among the Meso-American (the strip of land between North and South America) societies in astronomy. They were

Setting the Scene 9

FIG. 1.5 The Meso-American Zapotek pyramid on Monte Alban. (Courtesy of Matt Saunders, Creative Commons Attribution-Share Alike 3.0 Unported license.)

broadly aware of the cycles of the planets, and they especially studied the Sun, Moon and Venus, the morning and evening 'star.' Many buildings were aligned to act as observatories. It is possible that the Maya were the first to describe the Orion Nebula as being a blur, not a pinpoint star, before the advent of telescopes, and they could predict eclipses and transits.

The Mayan writing was, possibly, the earliest fully formed script that contained over 1000 glyphs—individual meaningful writing marks. The Mayan community at its peak of development spanned from A.D. 250–900, after which it collapsed for as yet unknown reasons. Perhaps, sadly, they are most famous for predicting the end of the world to be about December 21, 2012, which has been clearly proven wrong—you are reading this after that date! Proponents of the prediction claim its failure was because of confusion over the projections, assumptions and misunderstandings of ancient American texts.

The oldest known civilization on the South American continent is the Caral Supe (3000–2500 B.C.), located off the coast of Peru. And there were others, such as the Olmec (1200–400 B.C.) who constructed the first pyramids in the Americas; the Zapotec (500 B.C. to A.D. 750), known for its astronomical observatory building, the Nazca civilization (A.D. 1–700), famous for being (possibly) connected to huge geoglyphs (motifs in stone) found in the area, and other later ones such as the Incas (A.D. 1250–1532) and the Mississippian horticulturalists (A.D.1100–1450). Figure 1.5 shows the Meso-American Zapotek pyramid on Monte Alban.

Ancient Rome goes back to at least the eighth century B.C., when it began to develop across a trade and traffic route by the river Tiber in Italy. A significant contribution to observational science and astronomy of the Romans was their development of the calendar, modifying that previously used by the Mesopotamian civilization. Claudius Ptolemy (A.D. 90–168) was a Roman citizen of Egypt who wrote in Greek. He is most noted in astronomy for his huge treatise *The Almagest*, that was very useful in many ways, but he is infamous for holding back progress by insisting on his geocentric model for the Sun, Moon and planets.

The Development of Scientific and Astronomical Instruments

To say it was a surprise would be the understatement of the millennium! In 1900–1901 underwater divers were searching for sponges off the coast of a small Greek island, Antikythera, which sits between two other islands, Kythera and Crete. They found, along with sponges, artifacts from a sunken wreck off Point Glyphadia. What they retrieved from the sunken cargo ship, possibly on its way to Rome after looting the island, staggered the scientific community—and the world!

Among the statues and pots a rock was noticed with a gear wheel embedded in it. This was believed to be a clock mechanism and was ignored until 1951 when an English physicist, Derek J de Solla Price, examined it and concluded it to be an astronomical device dating from 150–100 B.C. With its 72 gears, 3 dials, and many inscribed scales, it demonstrated skills of engineering miniaturization and complexity not seen until more than a millennium later.

Following intense interest and study, the device appears to be the first analog computer, the purpose of which was to calculate the positions of the Sun, Moon and possibly all of the five then known planets as well as to tell the date and predict eclipses. The dials are marked with the Greek signs of the zodiac and the days of the year (Egyptian calendar), even allowing for the additional day every four years. This was about a century before the additional days for leap years were inserted in the Julian calendar. It has been suggested that the instrument, dubbed the Antikythera mechanism, shown in Figs. 1.6 and 1.7, might have been created by the great ancient astronomer Hipparchus. The find poses many questions.

FIG. 1.6 Antikythera mechanism, front view. (Courtesy of Creative Commons Attribution-Share Alike 3.0 Unported license.)

FIG. 1.7 Antikythera mechanism, rear view, revealing some of the finer tooth gears. (Courtesy of Creative Commons Attribution-Share Alike 3.0 Unported license.)

The Babylonians are credited with the first astronomical observations and writing and there is plenty of evidence that there was an ever-increasing awareness of the planets and their periodicities. The Venus Tablet of Ammisaduqua (King Ammisaduqa, ca. 1582–1562 B.C.), written in cuneiform, refers to observations of the rising and setting of Venus. The ancient Egyptians precisely aligned their pyramids with stars. The Greeks of the time developed highly sophisticated mathematics and built simple models to discuss geocentric or heliocentric theories, and other civilizations were also on the same academic path. But there was nothing like the Antikythera mechanism, or at least nothing has been unearthed so far, although it would be unexpected if this advanced mechanism were the only one of its kind.

The geared rock and associated artifacts are housed in the Egyptian National Archaeological Museum. Many detailed models have been produced to enable its suggested full working mechanism to be viewed.

Poking a stick in the ground and watching the change of position of its shadow could be called an astronomical instrument, and probably happened a few thousand years B.C. But place the stick at an appropriate angle and it becomes a sundial, the earliest examples of which are probably the ancient Egyptian and Babylonian obelisks and shadow clocks. The Roman writer and architect Vitruvius (ca. 80–15 B.C.) listed dials and shadow clocks known at that time. The 'Vitruvian Man' is a famous drawing by Leonard de Vinci, made after studying Vitruvius's notes to illustrate the ideal proportions of the human body, shown in Fig. 1.8.

In antiquity, sundials were made of stone, limestone, marble, glass, wood, string, tortoise shell, ivory with iron nails, silver, gold, pewter, bronze and other metals. Leather was used to make the pouches that carried the smaller portable models. They were made all over the world including the Middle East, Mexico, Europe, Africa, Australia and America. The beauty and complexity of a seventeenth-century German ring dial (Fig. 1.9) made from ruby, gold and rock crystal and conserved in the British Museum, is illustrated with the description that reads "Horizontal compass dial; set in gold ring; shoulder and sides of bezel chased with enameled scrolls; oval locket bezel; lid set with ruby surrounded by five crystals; compass inside; marked as horizontal sundial; holes for insertion of string gnomon."

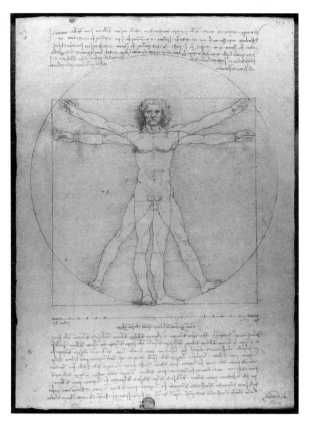

FIG. 1.8 Vitruvian Man, drawn by Leonard de Vinci to illustrate the ideal proportions of the human body. (Courtesy of Luc Viatour, http://en.wikipedia.org/wiki/File:Da_Vinci_Vitruve_Luc_Viatour.jpg.)

Sundials can be large landmarks or even memorials, such as the Kentucky Vietnam Veterans Memorial (Fig. 2.5). The gnomon, the stick bit that casts the shadow, is 14.62 ft above the surface and 24.27 ft long. As the shadow moves during the day it passes over the 1,103 Kentuckian names of the fallen. That must be a very emotional sight for many. Sundials may be associated with significant national history. An Australian dial, ca. 1837–1839, made by colonial engraver Raphael Clint and owned by the transported convict Daniel Cohen, has been purchased for placement in the Port Macquarie Historical Museum collection. It is a rare surviving metal object associated with penal settlement (Fig. 1.10).

Sundials come in all sorts of designs; horizontal, vertical, pocket, polar, reclining, spherical, cylindrical and more. Add a

14   Orrery

FIG. 1.9 Seventeenth-century German ring dial. (Courtesy of the Trustees of the British Museum, London.)

curved strip with a slit in the style part of the gnomon and the equatorial bow sundial begins to look like an armillary (Fig. 1.11). Sundials of the sixteenth to eighteenth century were particularly ornate, often made of brass, and were finely marked to determine the time, such as the two shown in Figs. 1.12 and 1.13. Figure 1.12 clearly shows the maker, R. White Fecit, and the date 1732, when dials were still being used to check the time on clocks and watches of the day, as continued for some time. Combined Sun and Moon

Setting the Scene 15

FIG. 1.10 Kentucky Vietnam Veterans Memorial. (Courtesy of Creative Commons, Public Domain.)

FIG. 1.11 Equatorial bow sundial. (Courtesy of Willy Leenders at nl.wikipedia, Creative Commons Attribution-Share Alike 3.0 Unported license.)

16   Orrery

FIG. 1.12 Eighteenth-century brass sundial. (Photograph by author with permission of the Science Museum, London.)

FIG. 1.13 Eighteenth-century brass sundial. (Photograph by author with permission of the Science Museum, London.)

dials have survived and reside in science museums. They incorporated not only local time, but also tides and other data.

Next to the sundial or sticks in the ground, an armillary is probably the most ancient astronomical instrument. It is a ball representing Earth with a band around that models the celestial sphere. The purpose was to demonstrate the movement of the stars around Earth. The instrument became more complex as knowledge of the heavens increased, and bands were added to represent the great circles of the ecliptic, meridians, colures and parallels when it was known as an armillary sphere. The Latin for rings is *armillae*, hence armillary. The ancient Greek astronomer Eratosthenes (276–195 B.C.) was possibly the inventor of the armillary. Eratosthenes is reputed to be the originator of the term "geography" and a system of latitude and longitude. He calculated the circumference of Earth and its distance from the Sun. His nickname was Beta since he was referred to as being second-best in almost every field.

The smaller armillaries were used to assist in observation or demonstration. Larger ones that contained much finer detail were used to study star positions in ecliptic or equatorial coordinates. Ptolemy described the earliest known of these complex instruments in *The Almagest* in 150 A.D. There was, in addition, a design by Renaissance and Islamic astronomers used to measure coordinates with respect to the celestial equator. Tycho Brahe (1546–1601), the Danish astronomer, built and used both types. Brahe constructed his own view of the heavens by combining the hitherto Ptolemaic view with the Copernican view, which was then dubbed the Tychonic system: the Sun, Moon and stars revolve around Earth and the five planets revolve around the Sun.

Many superb armillary examples exist, mostly in museums. The Istituto e Museo di Storia della Scienza in Florence, Italy, displays one made in 1557 by Girolamo della Volpaia that is gilded, 360 mm in diameter and mounted on a wooden base. Armillaries come in many sizes. An example in the same museum is very ornate and huge at 2,420 mm in diameter, constructed by Antonio Santucci around 1590 for Ferdinand the First.

A most impressive armillary in design and provenance is housed in the Oxford Museum of the History of Science in England. Unsigned and made about 1580, it is not as huge as the instrument in Florence, but is still only a little less than half the

18   Orrery

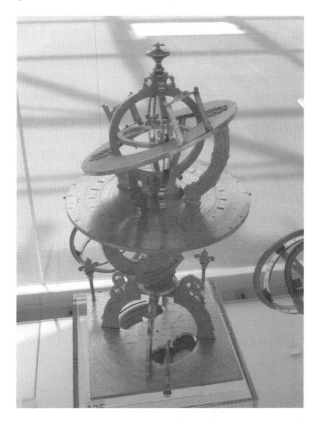

FIG. 1.14 The Austrian universal equinoctial armillary dial by Mathias Hauser and Bernard Polanski, 1750. (Photograph by the author, object displayed in the courtyard of the National Maritime Museum, Greenwich.)

diameter at 995 mm. The arms of Henry Percy, ninth Earl of Northumberland (1564–1632), are inscribed on the base to give it good provenance. Henry Percy was imprisoned in the Tower of London for 16 years, accused of involvement in the Gunpowder Plot to blow up the Houses of Parliament.

Museum Boerhaave in Leiden, the Netherlands, houses a gilt brass armillary dated around 1600 by Giovanni Paolo Ferreri, who also constructed three others that are preserved in the National Maritime Museum in Greenwich, England. An Austrian universal equinoctial armillary dial (Fig. 1.14) by Mathias Hauser and Bernard Polanski, 1750, sits in the courtyard of the same museum. The label reads: "This Austrian sundial works on the same basic principles as universal equinoctial ring dials, except that it can be

FIG. 1.15 The celestial sphere donated by the Woodrow Wilson Foundation, Ariana Park, Palais des Nations, Geneva, Switzerland, 2010. (Courtesy of Creative Commons CC0 1.0 Universal Public Domain Dedication.)

used to find both the time in your current location and in other cities around the world."

On the grounds of the United Nations headquarters in Geneva, Switzerland, stands a celestial sphere (armillary) placed there in 1939, donated by the Woodrow Wilson Foundation and sculpted by Paul Manship (1885–1966; Fig. 1.15). It is over 4 m in diameter and weighs 5,800 kg, with 85 gilded constellations represented and 840 silvery stars. However, due to the materials used and the design, it has deteriorated greatly from cracking and corrosion and is missing pieces.

The astrolabe, or planispheric astrolabe, is a two-dimensional model of the stars in the sky, the celestial sphere, with Earth at its center. Depending on the complexity of the model, it can be used to tell the time and simulate movements of the stars and other known astronomical objects for surveying and astrology. An Islamic fable has it that in the second century A.D., a celestial globe was dropped by the great astronomer Ptolemy when riding on his donkey. The donkey trod on it, flattened it and produced an image that Ptolemy decided would be useful to display the celestial sphere. He called it an astrolabe, holder of the stars.

It seems likely, however, that ancient astronomers were familiar with the astrolabe for a few centuries before this. The earliest surviving astrolabe is dated A.D. 927–928, and such instru-

FIG. 1.16 Eighteenth-century Persian astrolabe. (Courtesy of Andrew Dunn, Whipple Museum of The History of Science.)

ments were used until the seventeenth century. The earliest surviving treatise on the astrolabe was written by Theon of Alexandria, a Greek mathematician (ca. 335–405) well known for editing and arranging works by Euclid and Ptolemy. Theon had a brilliant daughter, Hypatia, who was the first woman to achieve renown in mathematics, astronomy and philosophy. In A.D. 415 she was brutally murdered by a mob whose hatred was probably whipped up by political and religious jealousy. A Hellenistic pagan, she was devoted to the study of magic, astrolabes and musical instruments. Early Islamic treatises from the ninth century, such as one by Messehalla, discussed astrolabes in great detail and greatly influenced the author of *The Canterbury Tales*, Geoffrey Chaucer (ca. 1343–1400), who also wrote a treatise on the astrolabe.

The planispheric astrolabe is built up from the celestial part, the 'rete,' the terrestrial parts, the 'plates,' a brass plate with a rim, the 'mater,' an index for the front, the 'rule' and another one for the back with additional sights, the 'alidade.' All were pinned together via a pin and wedge through a central hole. Different plates were used for different latitudes, although later designs were usable without having to change plates. The image in Fig. 1.16 is that of an eighteenth-century Persian astrolabe, where the points on the curved spikes indicate the brightest stars.

The list of types of instruments is huge and all were created in the most ornate and intricate manner. Fortunately, many examples survive in museums, especially from the fourteenth to seventeenth centuries, of compasses, dials (diptych, vertical, disks, chalices and more), celestial globes, folding rules, gunner's sights and levels, hodometers, theodolites, quadrants, protractors, measuring rods, mining instruments, primum mobiles, radio latinos, wind vanes, armillaries and more. The compendium contained several instruments in one that looked like a complicated and ornate flat-pack—quite possibly a status symbol to produce out of the pocket in those days! A brief description of all these instruments is given in the glossary.

All of the scientific and astronomical instruments of antiquity described have one thing in common. They all involved knowing, measuring, using or displaying the time. So it is not unexpected that philosophers invested large portions of their efforts in developing time instruments or clocks. Clocks are therefore scientific instruments in their own right, and, as described later, they became inextricably linked with orreries. John Wheeler (1911–2008) once quipped that time is what nature uses to prevent everything happening at once, and space is what prevents everything happening to me!

Depending on which reference is consulted, the word "clock" came from either the Sdcaon clugga, the Teutonic glocke, Latin 'clocca' or French 'cloche,' but all meant 'bell.'

How was their movement powered? "By clockwork!" Behind these familiar words is an exciting history, not just interacting wheels and cogs, but discovery, death, intrigue and lives sacrificially dedicated to unraveling the past and, against all odds, developing life-saving precision instruments.

Cogs and gears were not unknown in the ancient world, but sometimes they were large clanking monsters. To measure distance traveled, a type of odometer was developed in the Jin Dynasty (A.D. 265–420) in China. It consisted of a wooden figure connected to the wheels by a series of gears that beat a drum to indicate the distance covered. The circumference of the wheels was

5 m, and the figure would strike the drum every 100 revolutions, indicating that the carriage had traveled 500 m.

However, water clocks of the simplest type have been known for several thousand years. The oldest dates to between 1417 and 1379 B.C. during the reign of Amenhotep III, where one was used in a temple at Karnak. Water was allowed to drip from a stone vessel with sloping sides (to produce a constant flow) into a collector with water level lines to mark the passage of hours. The times and dates of religious and seasonal events could then be known in addition to the requirements of daily routines. Babylonian clocks used the weight of water rather than the volume. There is also evidence of water clocks in ancient Iran, India and other regions. Clocks based on the outflow of water were called "clepsydra" and were frequently calibrated for accuracy by comparison with sundials.

Water clocks from a few centuries A.D. began to use gears. The most notable of the period was the tower clock (Fig. 1.17) of the Chinese scientist Su Song (A.D. 1020–1101). The tower was 9.1 m tall, possessed a bronze power-driven armillary for astronomical observations, an automatically rotating celestial globe and five front panels with doors that allowed viewing of changing manikins that rang bells or gongs and held tablets indicating the hour or other special times of the day. Amazingly, the tower incorporated an escapement mechanism previously invented by Yi Xing and Liang Lingzan in 725 A.D. Escapement mechanisms will be discussed later but it is worthy of note here that in the Greco-Roman era the third century B.C. engineer, Philo of Byzantium, referred to escapement mechanisms already being incorporated into water clocks.

The use of water clocks knew no bounds. The Roman engineer Vitruvius mentioned alarm clocks, visits to Athenian brothels were timed by the clepsydra and the Hellenistic physician Herophilos measured his patients' pulse beats with one.

Only a brief overview of water clocks is given here, as there is a vast amount of information available on the Internet, in books and at museums on the use and development of these fascinating instruments that were common throughout the world for many centuries. The most sophisticated water clocks had come a long way since the ancient shadow clocks and the primitive Chinese

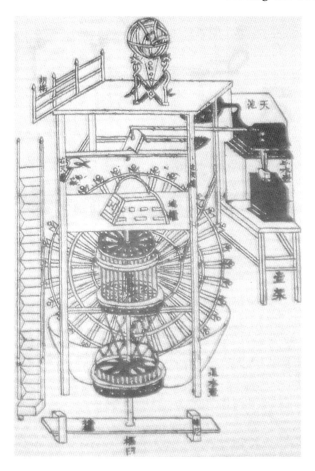

FIG. 1.17 The tower clock of the Chinese scientist Su Song. (Illustration by Su Song, printed in *Xin Yi Xiang Fa Yao*, 1092, reprinted in Joseph Needham, *Science and Civilization in China: Volume 4, Part 2, Mechanical Engineering.*)

and Japanese timekeepers, such as the kind with a wick 2 ft long that was processed to smolder rather than ignite, and time passed by observing the progress from knot to knot.

Mechanical Clocks

If the history of the orrery is forever tied to George Graham, then the history of George Graham is certainly the history of mechanical clocks.

Fig. 1.18 Lantern clock in Knole House, Kent. Possibly Cromwellian. (Courtesy of the National Trust, House and Collections at Knole.)

Of course, the various kinds of time-indicator instruments from ancient shadow sticks, dials and candle clocks through to highly complex water clocks were magnificent for their time and in their maker's ingenuity, design and usefulness. However, a major step forward in terms of accuracy was taken with the mechanization of devices in the fourteenth century, possibly even the thirteenth. The skills of the English clockmakers were really not fully developed at this time and the designers were blacksmiths by trade and training, often making simple lantern clocks. A well-preserved example of a lantern clock is to be found in Knole House (Fig. 1.18), where Charles Boyle spent his early school years. As for many timepieces of the age, the most accurate were correct to

Fig. 1.19 Astronomical clock at Hampton Court. (Courtesy of Creative Commons Attribution-Share Alike 3.0 Unported license. Attributrion; WazzaMan at en.wikipedia.)

within a quarter of an hour and that was considered good enough for anyone to regulate their affairs, unlike today, when seconds and even milliseconds often count to add to the pressure of modern living and doing business.

Highly ornate clocks and jeweled watches that were listed in Henry VIII's and Elizabeth I's wardrobe were likely bought abroad or made in England by foreigners. Even these items were probably status symbols or decorative and of little use to tell the exact time. They could be corrected daily by a sundial if they had one and if the Sun was shining!

Since continental clockmakers of the age were ahead of the game, Henry VIII brought in six of them to work on royal clocks. He was particularly keen to have a superb clock for Hampton Court, and the design is credited to the Bavarian astronomer Nicholas Kratzer (1487–1550), who was brought in by Cardinal Wolsey. The brilliant French clockmaker Nicholas Oursian (or Urseau) built the clock (Fig. 1.19). He also remained horologist for royal families until his death in 1590. The clock was an astronomical clock of great complexity. The 15-foot diameter dial consisted of three concentric copper dials rotating at different speeds. The clock indicated the hour, day, month, position of the Sun, signs of the zodiac, day number in the year, Moon phases and its age in days. Importantly for Henry, it presented the hour when the Moon crossed the meridian, thus the time of high water at London Bridge, which he needed to know as he often traveled by royal barge. Since this was prior to Copernicus, Earth is at the center of the clock, with the Sun going around it.

As the innovative skills of the heavy-duty blacksmiths waned it was the locksmiths who took over the finer developments such as decorating the fine clock encasements with engraving and enameling and precision work. But this was still mainly the province of continental artificers, although English clockmakers learned from them, and learned very well.

The French influence was still apparent in those early days, as Frenchman Nicholas Oursian became clockmaker to Edward VI, Queen Mary and Queen Elizabeth, although Oursian worked in London for most of his life. His son succeeded him as Royal Clockmaker. Then Bartholomew Newsam took over in 1590. He resided in the Strand near Somerset House until his death in 1593.

In the early seventeenth century a windmill stood in front of Somerset House, which was used to pump water. A maypole was re-erected there in 1661, 134 ft high, after being demolished by the Puritans. This mammoth task was said to be so difficult that Prince James, Duke of York and Lord Admiral of England, enrolled 12 seamen to bring their cables and other tackle, including six great anchors, to achieve the task. It was finally removed in 1717 and presented to, or purchased from, the parishioners by Sir Isaac Newton, who gave it to the Rev. Mr. Pound at Wanstead Park for the support of what was then the largest telescope in Europe, being 125 ft in length. (James Bradley, the third Astronomer Royal, was trained by his uncle, the Rev. James Pound.) Huygens constructed the telescope on the ex-maypole, using one of his three object glasses, with a huge 123-foot focal length. The telescope was subsequently donated to the Royal Society.

The magic ingredient for the advancement of accuracy was the escapement, a mechanism for the control of the rate of 'ticking,' or wheel movements leading to the turning of the hands. In the earliest models, this mechanism was used only to ring bells or gongs to announce religious or other important times during the day.

The principle of the regulated control of movement was described as early as the third century B.C. by Philo of Byzantium in

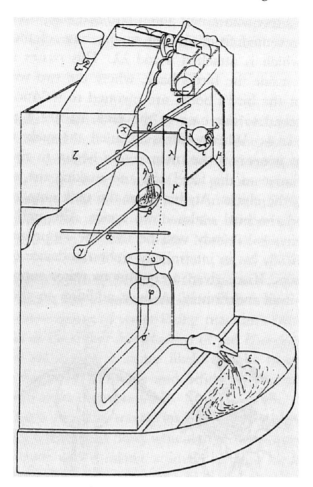

FIG. 1.20 Regulation of the flow of water in a third-century B.C. washstand. (Carra de Vaux, B. (1903): *Academie des inscriptions et des belles lettres: Notices et extraits des mss. de la Bibliothèque Nationale*, Paris 38, 27–235 (163))

his design of a washstand (Fig. 1.20). Water dripped from a tank onto a counterweighted spoon. When heavy enough, the spoon tipped the water into a wash basin while also releasing a sphere of pumice stone. With the weight of the water removed, the spoon reverted back to its original position, thus closing the door on the next piece of pumice. At the time, Philo referred to the mechanism's similarity to that of clocks that were clearly known to be already using the method.

28   Orrery

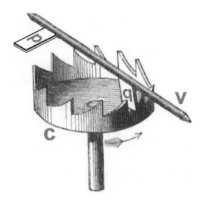

FIG. 1.21 Verge-type of escapement, where the paddles p and q alternately stop and release the wheel

FIG. 1.22 Foliot-weighted balance beam that controls the rate of rotation of the verge shown in Fig. 1.21

It was in the thirteenth century that the development of the escapement useful for tracking time began. The full mechanism (known as the 'movement') consisted of a weight, or spring, to act as the source of energy; the verge (escapement; Fig. 1.21) to control the action; and the foliot (Fig. 1.22) to smooth out the action.

Although this movement was a great leap forward, the resulting clocks were not all that accurate and may still have been half an hour or so off each day. They were often reset by referring to sundials. The big advance was the replacement of the weakest link, the foliot, by the balance wheel, which was much less susceptible to the vagaries of a swinging beam. The source of energy

Fig. 1.23 A fusée mechanism to compensate for the gradual weakening of an unwinding spring. (Courtesy of Creative Commons Attribution-Share Alike 3.0 Unported license. Attribution; Timwether.)

was a large weight, a huge stone, that was attached to a chain wound around a drum. As the drum rotated under the turning force of the weight, it transferred the movement to the axle or arbor of the escapement wheel via a system of cogs. However, someone had the unenviable task of having to regularly—usually at least daily—winding the weight back up to the top of the tower.

As an alternative to the dropping weight providing the energy, coiled springs were used in clocks as early as the sixteenth century. The problem with springs is that as they unwind to drive the movement, they lose their torque, or turning power. To compensate for the gradual loss before rewinding, a fusée (Fig. 1.23) was invented that was a cone-shaped pulley with a spiral groove around it and wound with a cord or chain attached to the mainspring barrel. As the clock runs, the chain is unwound from the

30 Orrery

Fig. 1.24 A skeleton clock with three fusées. (Photograph by the author with permission of the wardens of Derrynane House, West Ireland.)

fusée from the top to the bottom and onto the barrel, where the progressively weaker pull finds it easier to turn the larger radius and hence delivers a constant force. The system is easier to understand from the image. A skeleton clock (Fig. 1.24) exhibited in the Derrynane House museum, Kerry, West Ireland, employs three fusées to cope with the additional functions required for chiming the quarter hours. It was donated to Daniel O'Connell the libertarian in the early nineteenth century, but it may well have been constructed a long while before. Although the fusée was a great idea and lasted for a couple of centuries, it was cumbersome, difficult to maintain and expensive, so it was eventually consigned to history following further inventions to deliver constant energy. In fact, the fusée (often, and probably incorrectly, credited to Jacob Zech of Prague) was known prior to its use in clocks. Even Leonardo da Vinci (1452–1519) used it in one of his machines. The Germanische Nationalmuseum in Nürnberg, Germany, shows a clock that was built around 1430 and incorporates a fusée.

Possibly the earliest escapement clock was one designed by Richard of Wallingford (1292–1336), who was an English mathematician, astronomer/astrologer and horologist while serving as abbot of St. Alban's Abbey. However, he died during construction, and the clock was completed about 20 years after his death by William of Walsham. During the reformation under Henry VIII the clock was destroyed as part of the dissolution of St. Alban's Abbey in 1539. The escapement was not, in fact, a verge, but a strobe consisting of two escape wheels on the same axle with alternating radial teeth.

A very complex astrarium clock, based heavily on the astrolabe for indicating the Ptolemaic movements of sky objects and calculating astronomical events, was created by the Italian physician and astronomer Giovanni de Dondi (1318–1389). The device was put together without any screws, just tapered pinions, and included 7 dials and 107 gear wheels. Unfortunately, it was destroyed, but fortunately, Dondi recorded every detail of the instrument for later reconstruction. To indicate dimensions in his descriptions, de Dondi used units such as the width of a goose quill, the thickness of a blade of a knife or the breadth of a man's thumb.

Some of the earliest mechanical clocks were a group of astronomical clocks in cathedrals in the southwest of England, namely Salisbury, Wells, Ottery-St.-Mary, Exeter (the main dial is the oldest part and dates from 1484) and Wimborne. Many of the facts concerning these clocks are 'probables' and informed guesses, although it appears likely that Salisbury Cathedral clock was constructed before the bishop moved to Wells, bringing the clockmakers with him. The Salisbury Cathedral clock is a large, iron-framed clock without a dial and is located in the aisle of the cathedral. It probably dates from about 1386. The Wells clock was 'updated' in the seventeenth century by converting to a pendulum and anchor escapement and was subsequently installed in the Science Museum in 1884. Some of the parts may even have originated from Glastonbury Abbey.

Note that for a little less than a hundred years the Wells clock was manually wound by five generations of the Fisher family, who turned the three 250-kg weights 800 times three times per week. Recently, Paul Fisher announced his retirement, and the mechanism was replaced with an electric motor.

A wonderful and beautifully restored medieval astronomical clock (Fig. 1.25) can be seen by anyone standing in the center

32   Orrery

FIG. 1.25 Prague astronomical clock, or Orloj. (Courtesy of Irene Hon.)

of Prague, the capital city of the Czech Republic. It was first installed in 1410, making it the third-oldest astronomical clock in the world. The clock, or Orloj, is mounted on the wall of the Old Town City Hall in the Old Town Square. It comprises three components: the astronomical dial representing the positions of the Sun and Moon in the sky; a clockwork hourly show of the Apostles, with other moving figures such as a skeleton representing death; and a calendar dial.

In France, the so-called Great Clock (Le Gros Horloge) was installed above a bridge spanning a street in Rouen in 1389. It has had some additions and the foliot was replaced by a pendulum, but it was the first clock to strike the quarter-hours.

In the Japanese Edo era (1603–1868), many highly ornamental clocks were crafted, called *wadokei* (literally 'Japanese clocks'). They were unique in that in the old Japanese time system, time was not measured in equal units (such as hours) as it was in Western countries. A day was divided into the daytime and the nighttime, in accordance with sunrise and sunset, and each was further divided into six segments. Time corresponded to the position of the Sun, but the six daytime segments were not the same length as the six nighttime segments (other than at the equinoxes). Those segment lengths, moreover, changed slightly every season. The Edo era saw many master clockmakers. Famous masters included Sukezaemon Tsuda, in the early Edo era; Kichirozaemon Kono and Riemon Hirota, in mid-Edo; and Hisashige Tanaka (commonly known as "Karakuri" Giemon) and Norichika Ohno, all working for the government to produce *wadokei*.

Of course, clocks have often been a quality gift, not least among royalty and nobility. The Anne Boleyn clock, now in the library at Windsor Castle, is said to have been a wedding gift from Henry VIII to Anne on their wedding day in 1532. The two weights are engraved with 'H' and 'A' and true lovers' knots, and the clock and ornate copper guilt case stand on a bracket. Queen Elizabeth I was gifted with many clocks and watches, not only by her clockmakers and keepers, but notably by the Earl of Leicester and Lord Russell—diamond and ruby encrusted ones, of course.

Suspicions of Guy Fawke's planning of the plot to blow up the London Houses of Parliament, with King James I, when arrested were confirmed as he had upon his person a watch procured for the purpose of ascertaining the exact time of burning of the touchwood till it reached the grains of gunpowder. Watches were not common around that date of 1605.

An enormous leap toward increased accuracy was made in the mid-seventeeth century through the replacement of the foliot by a pendulum, at which time such clocks were called regulators, because they were used to regulate (check the time of) other clocks. The basis of pendulum accuracy was that the period of swing for a given length was the same whatever the angle or displacement of swing. A 'seconds pendulum' was constructed using a pendulum 0.994 m (39.1 inches) in length with a swing of exactly one second, varying slightly, of course, with environmental conditions and position on Earth.

Instead of having large swings of around 100° that could, in fact, be somewhat unstable, Christiaan Huygens (1629–1695) re-

Fig. 1.26 A Mezzotint portrait of Christiaan Huygens around 1670. (Courtesy of the Trustees of the British Museum.)

duced that angle to around 5° to give a more reliable periodicity with a consequent advantage of not only accuracy, but a more manageable size of the clock case (Fig. 1.26). Huygens is thus recognized as the inventor of the first practical pendulum clock in 1656, taking advantage of Galileo's (1564–1642) observation of the properties of a swinging pendulum. In 1657, Huygens gave the job of production of his clocks to Salomon Coster, a master clockmaker living in The Hague. This is the same Huygens who studied the nature of the rings of Saturn and discovered Saturn's moon Titan.

Of course, everyone wanted a piece of this action, and John Fromanteel, a clockmaker of Dutch descent, traveled to Holland to work with Coster to learn the pendulum construction commonly labeled Huygen's 'pendule.' John, son of Ahasuerus Fromanteel

FIG. 1.27 Dial of Fromanteel eight-day, weight-driven, long-case clock. (Courtesy of the Trustees of the British Museum.)

(1607–1693) and a member of the Blacksmiths and Clockmakers Companies, was an obvious choice to work with Coster, as he also had a business in Amsterdam, where his family originated.

It is not surprising that such exciting inventions would spread rapidly, and by 1658 Fromanteel was selling the wonders in London. Figure 1.27 shows the dial of a Fromanteel eight-day, weight-driven clock whose case was ebonized pear-wood-veneered oak. Those who could not afford the full works had their old clocks upgraded to include a pendule. Miniaturization of the movement to enable the production of small clocks or watches was possible due to the foliot and pendulum being replaced by a balance wheel instead of a pendulum. Very significant advances were subsequently made to the escapement mechanism, as will be described later.

The pendulum seemed to mark a change in direction for clock production. No longer was it the blacksmiths copying continental mechanisms. It was the birth of English horology under the encouraging reign of Charles II.

As for many of the late-seventeenth and early-eighteenth century clockmakers, George Graham owed much to the master of clocks often referred to as the father of English watchmaking, Thomas Tompion. Very little is known of Tompion's early years, although his ancestors may have been French and sold bungs and plugs for barrels and guns. There is a naval saying that you only fire one round with the Tompion in. His great-uncle was probably James Tompyan, an editor of the Bedfordshire Parish registers and resident of Southill, neighboring Northill. Tompion was born in 1639 in Northill parish, Bedfordshire, and probably worked as a blacksmith with his father, also Thomas Tompion, until 1664, when he became an apprentice to a London clockmaker, or so it is said, since there are no records to indicate to whom he was apprenticed.

Maybe, as an anecdote tells, he became competent in the art through mending a clock for a gentleman and subsequently believing, and proving, that he could make one of his own! In fact he joined the Worshipful Company of Clockmakers of London in 1671 as a brother and was admitted as 'a Free Clockmaker upon Redemption,' that is, paying a fee of ten shillings, thus bypassing the necessity of having served as an apprentice. He was described as a Master Blacksmith-clockmaker, specializing in the making of large turret or church iron clocks. He became a Master in 1704, just a few years before his death in 1713.

There were many such worshipful companies at the time, and they sought to impose high standards of workmanship, ethics and behavior. Oaths had to be made, and anyone who wished to pursue a craft needed to belong to such a company, as it was forbidden for those outside the company to practice. They also had to be Freemen of the City. Many members or liverymen of the Clockmakers Company were blacksmiths, their craft being more defined by general metal working than by the shoeing of horses, which was more the job of the farrier. Therefore they were already freemen of the Worshipful Company of Blacksmiths and came with good pedigree, probably having been working with the ferrous metals involved in the production of tower and church clocks. A bell, cast for the church of St. Lawrence at Willington near Northill, was inscribed "Thomas Tompion Fecit 1671." Domestic clocks and watches were mostly imported, but many of the 'aristocratic quality' ones, such as Tompion's gilded Masonic watch shown in Fig. 1.28, were not.

Fig. 1.28 Thomas Tompion gilt-brass and leather-cased verge watch with alarm, around 1670. (Courtesy of the Trustees of the British Museum.)

Most companies, or guilds, had religious affiliations, and so had turbulent histories not least because of the behavior of monarchs. The Blacksmiths' Company suffered greatly under Henry VIII and the dissolution of the monasteries and also subsequent civil wars, but still managed to survive to the present day. London now has over 100 "companies." Contrary to a common guess because of the original use of the word 'guild', the companies were/are not connected to freemasonry, although both have always donated huge sums of money to charitable causes.

The importance of being a 'proper member' of a guild is exemplified by the story of another great and illustrious clockmaker who was alive at almost the same time as Tompion. Joseph Knibb (1640–1711) was a clockmaker in the city of Oxford, but freemen objected to him practicing there because he was not a freeman and demanded that he "suddenly shut his windows." But Trinity

College appointed him, officially, as the gardener, and so he was able to continue making his clocks as he was officially a gardener living on the premises but actually clockmaking in an undercover way. However, he was eventually admitted to the freedom on payment of a fine of £6.13s.4d and a leather bucket! He went on to be one of the finest and most inventive clockmakers of his time, coming up with the system of Roman striking and the tic-tac escapement. Eventually he was appointed clockmaker to King Charles II of England, then to King James II. From leather bucket to royal appointment!

Joseph Knibb was likely the Collector of Rates for Fleet Street at the time of Tompion, moving from Water Lane to the corner of Fleet Street and Water Lane, around 1675. Currently, the earliest record of Tompion in Water Lane, London, is in a tax collector's notes of 1671.

When Joseph became a member of the London Clockmakers' Company in 1670, Knibb's younger brother John had to look after the clockmaking business and also had some difficulties applying for his freedom, since he had none of the necessary qualifications. With the help of the mayor, William Cornish, and the MP for Oxford, Brome Whorwood, he eventually became a freeman in 1673 on payment of a fine of £20, reduced from the suggested £30. Generations of the Knibb family were not only among the very top clockmakers of the time (the Clockmakers of Claydon) but some became mayors of Oxford. Recently, a seventeenth-century clock by a 'former mayor' was sold at auction for £ 75,000, almost three times its estimated price. Many clockmakers continued to produce sundials and particularly small or pocket dials, often purchased as status symbols. Figure 1.29 shows an engraved silver dial made by Tompion between 1700 and 1710.

Tompion was revered in his field and many sought to learn their trade with him. His many apprentices included George Allett (originally apprenticed in 1863 to Solomon Bouquet, but joined Tompion in 1691), Edward Banger, Henry Carlowe, Daniel Delander (apprentice or servant to Tompion?), Ricard Ems, Ambrose Gardner, William Graham (nephew of George Graham), George Harrison (originally apprentice to Johana May, 1689, but joined Tompion in 1698), Whitestone Littlemore (originally apprentice to Thomas Gibbs), Jeremiah Martin (previously apprentice to Wil-

FIG. 1.29 Thomas Tompion horizontal inclining dial, silver; pierced and engraved folding gnomon with erectable extension. (Courtesy of the Trustees of the British Museum.)

liam Dent), Charles Molins (Molens?), William Mourlay, Charles Murray, Robert Pattison, William Sherwood (previously apprentice to James Delander), Richard Street (maker of a clock that was presented in 1708 by Sir Isaac Newton to Dr. Bentley, Master of Trinity College, Cambridge), Charles Sypson, William Thompson, James Tunn and Thomas White, many of whom became important clockmakers in their own right.

Whenever Tompion clocks come up for auction, they command huge prices. In a sale event for Sotheby's in 1999, "Masterpieces of the Time Museum," a 1705 Tompion quarter-hour chiming table clock sold for over $2 million. Only three such clocks are known, and all feature a red tortoiseshell exterior, gilt-brass mount and 'grande sonnerie' chimes. It was a world record for Tompion and English clocks. But even that huge price was dwarfed by the *world's* most expensive, the Duc D'Orleans Sympathetic clock, made about 1795, that realized $5,777,500 in the same sale. The maker was the renowned Abraham-Louis Breguet, (1747–1823), watchmaker to Louis XVI, Marie Antoinette, Napoleon and the sultan of the Ottoman Empire. A removable pocket watch was included within that was wound and the time set (in sync with the clock) by a mechanism in the clock's movement. Figure 1.30 shows a Breguet astronomer's clock with a viewing window to see the escapement.

40   Orrery

FIG. 1.30 Abraham Louis Breguet Astronomer's follower (observing) clock. (Courtesy of the Trustees of the British Museum)

Breguet almost lost his head in the French Revolution, but was saved from the guillotine by Jean-Paul Marat, one of the leaders of the revolt. It was only fair, though, as Breguet had previously saved Marat from an angry crowd!

As another indicator of Tompion's brilliance and standing in the scientific society, he was one of only a few watchmakers to become a member of the prestigious Royal Society. It was also an honor for him that when the Royal Observatory was established on June 22, 1675, King Charles II chose him to create two clocks based on an escapement designed by Richard Towneley, which would only require winding once a year. The clocks were very accurate and of great assistance to astronomers.

Tompion certainly moved in the right circles such as with the first Astronomer Royal, John Flamsteed, and the noted scientist Robert Hooke, who assisted him to make some of the first watches with balance springs that were much more accurate than earlier watches. He invented the first widely used balance spring regulator for pocket watches that was used until the late nineteenth century and the cylinder escapement that allowed him to create flat watches.

A significant record-keeping advance that Tompion introduced was the serial numbering system applied to his creations and thought to be the first time such a system was used for manufactured goods. It was just as well to have a method of keeping track since he produced about 5,500 watches and 650 clocks during his working life, somewhat due to his being the first to apply a systematic method of production to watch- and clock-making without any deterioration of quality. Indeed, the standards were so high that parts between watches were interchangeable. Upon the reverse of the British monetary bank note is currently a portrait of Adam Smith entitled "The division of labor in pin manufacturing." One might glean from this that Smith was the discoverer or developer of such a practice. Not so! He may have further focused on observing the qualitative improvements and applied it to a larger scale, but there were many practical examples before him, not least by the ancient Sumerians. Tompion developed his own such innovation where advantages accrued from division of labor. Sir William Petty, a political economist of the seventeeth century and familiar with Tompion's output, described the making of a watch, "If one Man shall make the Wheels, another the Spring, another shall Engrave the Dial-plate, and another shall make the Cases, then the Watch will be better and cheaper, than if the whole Work be put upon any one Man."

Among the 350 clocks and watches that are held in the Queen's residence, Buckingham Palace, two of Tompion's one-year clocks are still operational today, as are many of his other timepieces. Figure 1.31 shows a beautiful oak and 'mulberry' wood, long-case, one-month-duration movement made in London by Tompion between 1695 and 1705. In fact, although the clock is referred to as the mulberry clock, research shows that it is a burr wood, probably maple treated with nitric acid followed by linseed oil and lampblack to bring out the figures in the grain.

Richard Towneley (1629–1707), who worked with Tompion, was a clock enthusiast and also a mathematician and astronomer. In fact, he was one of a renowned group of astronomers that included Jeremiah Horrocks (the first to predict, and one of two to observe, the transit of Venus in 1639), who died young, at about age 24, William Crabtree (the other person to observe the Venus transit) and William Cascoigne (who invented the micrometer that Robert Hooke used to measure the size of comets and other cosmological bodies). The tercentenary anniversary of the beginning of rainfall recording in the British Isles, started by Richard Towneley, was celebrated in 1977 by British meteorologists.

In answer to a request by John Flamsteed to help with proving that Earth rotated at a constant speed, Towneley designed a new clock escapement mechanism, the deadbeat, to prevent the jerking movement of the second hand of pendulum clocks. Using money provided by Towneley's friend, Sir Jonas Moore, Surveyor General of the King's Ordnance, Tompion was commissioned to build two such astronomical clocks that were installed in the Greenwich Observatory on July 7, 1676. The clocks were difficult to maintain, however, and Tompion improved upon the mechanism, but it was only around 1715 that George Graham created a deadbeat that was truly successful. Towneley played a major part in the formulation of Boyle's Law, relating the pressure and volume of a gas, but more on that revelation later. There were so many more brilliant clockmakers of the seventeenth and eighteenth centuries, such as Daniel Quare and Nathaniel Barrow, that only a few could be mentioned here, but among the very best was George Graham, as we'll see later.

It is worth mentioning here that there were other brilliant clockmakers on the continent, such as Jean Vallier, who produced the stunning watch in 1630 shown in Fig. 1.32. The dial is of the highest quality. A small rectangular aperture at the top of the dial indicates the date above a subsidiary dial indicating the age and

Fig. 1.31 Thomas Tompion fine wood long-case clock known as the 'Mulberry clock' although it is not actually mulberry wood. (Courtesy of the Trustees of the British Museum.)

Fig. 1.32 Clock watch by Jean Vallier around 1630, Lyons, France. (Courtesy of the Trustees of the British Museum.)

phase of the Moon. To the left another dial shows the days of the week, with each day's ruling deity appearing in a sector. On the right the quarters are indicated on an engraved silver chapter ring. The bottom dial is for the hours, with an alarm-setting disc at its center. Two sectors in the upper right-hand area show the seasons and the months, with the number of days in each.

Not a great deal is known of Tompion's domestic life, although he appears not to have married, and his health was not always good. Possibly for that reason, he never finished making a clock that would go for one hundred years without winding, for the St. Paul's cathedral building in progress. Instead, after giving thirty years of dedication to clocks, he spent more time in relaxation and longer periods away from London. In particular he

Fig. 1.33 Portrait of Thomas Thompion. (Courtesy of the Trustees of the British Museum.)

spent much time visiting Bath, possibly to make use of the healing properties of the hot mineral waters. Tompion joined in partnership with George Graham in 1711, and a plaque hangs in Fleet Street, London, to commemorate both giants of the horological fraternity. He died in 1713 and is buried in Westminster Abbey. Figure 1.33 shows a portrait of Thomas Tompion.

Although little is known with regard to the minutiae of Tompion's daily existence, pertinent extracts from the diary of the great and renowned experimentalist, Dr. Robert Hooke, reveal some moments. A plethora of books and references have been created to record and describe the works and contributions to science of this seventeenth-century genius, and it is not necessary to repeat them all here, save as to mention the terse nature of some of the entries. Hooke was not a 'people person.' He liked things and objects. He became obsessed with the idea that everyone was stealing his ideas—and some definitely were. He was constantly falling out with those he had to work with, such as Wren, Newton and Flamsteed. He was described by Richard Waller, F. R. S., as appearing a bit of a despicable creature; "crooked, meager aspect, thin nose, hair long and neglected, went stooping, slept little and was melancholy, mistrustful and jealous"—and being "of stunted growth."

It is understandable, then, that the bitter Hooke wrote in his diaries the way he wanted to without any thought of recording for posterity or others to read. The writings were also a basis for remembering activities and costs for future use. His flippant entries and disregard for convention are illustrated in his references to Tompion, written also as Tomkins and Thomkin, not bothering to inquire as to the correct spelling or maybe deliberately demeaning the man with his loyal conversational companion, his diary, until he got to know him better.

Hooke was so furious when he heard of Huygen's pendulum watch (1675) that he said, with justification, that he had invented it earlier. However, Hooke didn't publish, and, in his annoyance, he did not accept terms for its production. He ignored the golden rule of inventions: "If you stop in your tracks, you will get run over!"

Hooke had demonstrated to members of the Royal Society, of which he was a leading light, the principles of his new invention called a quadrant, a new type of astronomical instrument. In 1673-74 he was asked by the society to employ a craftsman to build the quadrant, Tom Shortgrave, who would be paid £10. Hooke seemed to be slow getting on with the job, but the delay appears to have been because he was not happy with the abilities of Shortgrave.

Fortunately, Hooke then came across Tompion and entered in his diary, on April 5, 1674, "Began the description of quadrant," meaning he put pen to paper to create the design. On April 20 appeared "Cald on Tomkins for quadrant" and on April 28 was recorded "At Tomkins in Water Lane". On April 30 the society records that "Mr Hooke excused himself, that his quadrant formerly promised was not yet ready." Clearly, Tompion was taking longer than anticipated to complete the construction. On May 2, "To Thomkin in Water Lane."

As respect between the two great men grew, the diaries became a little more informative: "Much Discourse with him about watches. Told him the way of making an engine for finishing wheels, and a way how to make a dividing plate; about the torme of an arch; about another way of Teeth work; about pocket watches and many other things.". Having "Sent for quadrant from Tomkins" and a note to "...send him 25 sh" Hooke, on May 15, "presented the quadrant to Sir Jonas Moore."

The diary shows that Hooke eventually got the spelling right. "At Tompion in Water Lane," and showing great pride in Tompion by taking Flamsteed to see him at Water Lane, "I showed Flamsteed my quadrant." Flamsteed criticized the instrument, which made Hooke write when back at his abode in Gresham College, "He is a conceited cocks comb." The quadrant was still not finally completed when Hooke wrote on June 8, "Gave Tompions man 1sh". On July 5 the instrument was finished, and Tompion took it along to Gresham College to meet with Hooke and Jonas Moore: "tall and very fat, thin skin, faire, cleare grey eie." All went well, and Hooke wrote "Received home quadrant from Tompion. Sir J. More and Tompion here and at Blacklocks coffee house." On December 3 he gave a presentation of his quadrant to the Royal Society, giving full credit to Tompion:

> [B]ut if any person desire one of them to be made, without troubling himself to direct and oversee a Workman, he may imploy Mr Tompion, a Watchmaker, in Water-Lane near Fleet Street; this person I recommend, as having imploy'd him to make that which I have, whereby he hath seen and experienced the Difficulties that do occur therein; and finding him to be very careful and curious to observe and follow Directions, to compleat and perfect his work, so as to make it accurate and fit for use.

Likely due to Hooke's recommendations, Tompion was making another quadrant for the adventurer Francis Vernon, who unfortunately was killed in Persia over a penknife. Hooke: "At Mr Tompions about Vernon quadrant." The two began to meet more socially in the coffee houses; particularly mentioned are Joes (Mitre Court, Fleet Street, currently Messrs Hoare, Bankers; Fig. 1.34), Mans (Chancery Lane), Childs (St. Paul's Churchyard) and Garaways (Exchange Alley Cornhill). From June 27, 1674 to September 25, Hooke refers to conversations with Tompion at

FIG. 1.34 Mitre Court, Fleet Street, currently Messrs. Hoare, Bankers. (Photograph by the author.)

Joes, Garaways, Tompions house and Mans about a new dividing compass's screw upon a rule, clepshydra, plug for wind pump, the fabric of muscles, cork bladders, a fire engine, bellows, a barometer bolt-head and many other ideas. He also reported "I was very brisk smoked four pipes" and to drinking chocolate. As happened for most of Hooke's friends, he began to suspect them of pinching his ideas, in particular, Tompion and Robert Seignior the clockmaker—"To Garaways, Signor and Tompion cheats grand." Hooke and Tompion never fell out permanently, but Hooke, being somewhat neurotic, wrote in his diary entries such things as "Fel out with Tompion" and "Thompion. A clownish churlish Dog" when Tompion was actually supporting Hooke over the Huygen's pendulum watch.

Hooke appeared to have taken on the duty of advancing Tompion and introducing him and his talents to the world. Tompion's fame really took off in 1674, greatly due to Hooke.

How Tompion found the time to read contemporary technical journals as well as socialize and run his business and inventions is amazing. He and George Graham are listed as subscribers to John Harris's 1704 *Lexicon Technicum* ("An Universal English Dictionary of Arts and Sciences"). However, he declined to be too involved in the administrative duties of the Clockmaker's Guild, probably because they would have taken him away from his business.

Hooke's diaries are uniquely rich in the knowledge and understanding of events in Fleet Street (known as the Great Way from London to Westminster) and its surrounds as well as in the personal lives of clockmakers and scientists of the time. It appears that Tompion spent most of his money in the move to Fleet Street and needed a loan of £80 that Hooke would not lend, but who catalyzed a £50 loan from Sir Jonas Moore. Additional historical details are gained from many adverts Tompion placed in The London Gazette ffering rewards of several guineas for the finder of his valuable timepieces fallen from a gentleman's pocket."

Tompion's business flourished and he soon had many servants (probably apprentices) and garret-masters helping with piecework for the mass production of clocks and watches at the larger workshops. His sister, Margaret, and her two children also came to live at his house following the death of her husband Edward Kent. According to the rates assessment of July 16, 1695, the purpose of which was to raise money to fight France, there were 19 people listed as living at Tompion's residence. Nonresident clockmakers were also employed here, presumably to cope with the great demands for clocks and watches at home and even abroad. Tompion was so famous that the court painter, Kneller, painted his portrait, nobility carried his watches, counterfeit ones were appearing on the streets and Christopher Wren referred to Tompion making and servicing clocks at palaces. Reports suggest that Tompion was to construct a clock for the newly built St. Paul's Cathedral, but the commission was eventually given to Langley Bradley, possibly due to the large amount of money being asked by Tompion (Fig. 1.35).

Towards the end of the seventeenth century Tompion was pretty rich, which was certainly helpful when waiting for months or years for his bills to be settled. William III, on his death, owed Tompion £564.15s for watches and clocks given to the Duke of Florence, which William's successor, Queen Anne, refused to honor.

Even though robbery was rewarded by the ultimate penalty at the ropes of Tyburn, there were still many about that made paying and receiving money using portable cash highly risky. Luckily, then, this was the period of early banking and paper transactions. Goldsmiths took on the new service of banking, and Tompion

Fig. 1.35 As no contemporary portrait of Robert Hooke seems to have survived from the seventeenth century, this one is a reconstruction from the descriptions by his colleagues Aubrey and Waller. It shows him with a spring, pocket watch, fossil and map of the city of London after the Great Fire of 1666. (Courtesy of Rita Greer)

used the services of Sir Richard Hoare at the sign of the Golden Bottle at the end of Fleet Street near Temple Bar. Tompion gave cash to Hoare. Hoare gave Tompion a piece of paper. Tompion gave or sent his creditor the piece of paper that was exchanged for cash with Hoare. Such was the success of the piece of paper, the forerunner of the check, that records showed that Tompion had used it to part with almost £1,500 over a nine-year period.

Tompion's health deteriorated during his later years, and he spent much time at Bath and the Pump Rooms, seeking respite from his discomfort. He died on November 20, 1713. It is possible that Tompion spent time on some scientific inventions such as a diving bell, but there are no records of his to confirm this. From Hooke's diaries, Tompion was involved with making a barometer in 1676: "Cald on Tompion about new Barometer." Hooke reveals that he, with Mr. Boyle, Sir Jonas Moore (whom Hooke refers to as "dog More," once again the misspelling was characteristic of Hooke's additional attempts to demean those he disliked), Sir Samuel Moreland and Win (a local instrument maker) were, at times, with Tompion, which confirms the high society with which he mixed and the scientific work he undertook. It is probable that he collaborated with Flamsteed in a design for a decimal clock in 1691, according to some notes of Flamsteed. Three of his barometers are known and many dials.

Tompion was, and always will be, known for his mastery of clocks and watches and his nurturing of, and partnership with, George Graham.

2. Honest George, Chronometers and the Mystery of the Disappearing Proto-Orreries

George Graham, Clocks and Chronometers

To clockmakers George Graham was and is well known, even revered, although in this modern age his name does not commonly get a mention in scientific conversations and technology debates. But he was more than just a maker of clocks and of geared models that later became known as orreries. He was a scientist, passionate about astronomy, a socialist and worked with the best of the best. He was not just helpful to so many others but his generosity of spirit and largesse was absolutely crucial to the later successful introduction of a timepiece invention that saved so many lives at sea. No wonder that he earned the kindly nickname of Honest George.

Figure 2.1 shows Graham, born on July 7, 1673 (maybe 1675) in Horsgill, Cumberland, England. This may be Hethersgill, with a current population of around 400, in the city of Carlisle district in the parish of Kirklinton, Cumbria. Graham's parents were George and Isobel, who lived as peasant farmers and Quakers (although the Quaker approach to life was taken up more by his son than by him). The father died soon after his son's birth. George was brought up by his brother William (not a Quaker), who was also a good clockmaker and lived nearby in Sykeside.

At an age of only about 14 years, George walked to London, where he apprenticed himself to the clockmaker Henry Aske. After seven years of learning the trade with Aske, Graham must have attained great skills, as he was granted the freedom of the clock-

FIG. 2.1 George Graham. (Courtesy of the National Maritime Museum, Greenwich, UK)

maker's company. At this time he became assistant to Thomas Tompion, the leading clock-, watch- and instrument maker of the country. In 1704 he married Tompion's niece, Elizabeth Tompion, but they had no children.

Little is known of life within the Graham household, although one reference is quoted, but not substantiated, that the marriage may not have run smoothly. Elizabeth was reputed to have had two sons, whose legitimacy George refused to acknowledge. It is possible that the recorded falling out between Tompion and his associate Edward Banger was due to the accusation that Elizabeth's infidelity might have been with Banger—if, indeed, the rumors were true. Graham later became Tompion's business partner and was one of the executors of Tompion's will upon his death in 1713.

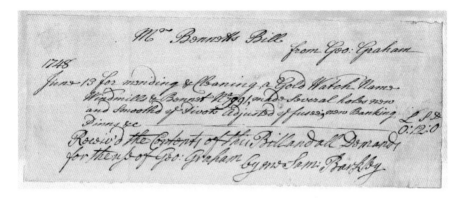

FIG. 2.2 George Graham's hand-written bill to a Mr. Bennett in 1748 for the mending and cleaning of a gold watch. (Courtesy of the Trustees of the British Museum.)

Tompion bequeathed his money as well as the land and property at Northhill, Bedfordshire, to his close relatives.

Not unexpectedly, Tompion bequeathed his business to Graham, who also, in conjunction with his wife, was a residual legatee. The business included "all of the stock and work, finished and unfinished and to continue the trade at Mr. Tompion's dwelling house at the sign of the Dial and Three Crowns, at the corner of Water Lane, Fleet Street, London, where all persons may be accommodated as formerly." Although likely keeping interest in the Tompion residence—probably through Obadiah Gardner, a former apprentice—Graham moved in 1720 across the road to the Dial and One Crown. This was a new house next door to the Duke of Marlborough's Head Tavern and the Globe. Graham remained there until his death in 1751. A hand-written bill to a Mr. Bennett in 1748 is for the mending and cleaning of a gold watch (Fig. 2.2).

The shop must have looked just like the painting on a modern-day box of chocolates—a quaint little store with two bay windows, one either side of the door. It remained with little alteration the home and premises of future watchmakers Mudge and Duttons for many years. Later it became the offices of the Sporting Life, and huge buildings now stand on the site. But there is still a Blue Plaque there to commemorate the dwelling place of Graham and Tompion.

Graham certainly worked closely with many great instrument makers. From 1720 to 1728 Thomas Wright occupied the shop, with Graham at the room upstairs, in which he remained until his death. Also in the immediate neighborhood were John Rowley and Richard and E. Cushee, two globe-makers at the sign of the Globe and Sun. Two former assistants to Rowley also worked there, John Coggs, Senior, and William Wyeth.

Graham was buried in the nave of Westminster Abbey with his former partner Tompion. The commemorative floor slab was taken up by the order of the Dean in 1838, at which time the London *Saturday Journal* of 1842 reported a quote that the men's memory will last although the slab was gone. It was replaced in 1866 by order of Dean Stanley. The Journal also reported that they were "greater benefactors to mankind than thousands whose sculptured urns and impudently emblazoned merits that never existed." The re-cut inscription on the slab is: "Here lies the body of Mr. Tho. Tompion who departed this life the 20th of November 1713 in the 75th year of his age. Also the body of George Graham of London watchmaker and F.R.S. whose curious inventions do honor to ye British genius whose accurate performances are ye standard of mechanical skill. He died ye XVI of November MDCCLI in the LXXVIII year of his age."

George Graham, "Honest George," was noted for many things in his lifetime, but a few achievements in particular can be highlighted—his work on the clock escapement mechanism, his association with The Royal Observatory at Greenwich, a mechanical tellurian, later to be dubbed an orrery, and generous support given to John Harrison, which was essential in solving the 'longitude problem.' He also invented scientific instruments other than clockworks.

The escapement mechanism that helped drive the train or gears that led to the time or event display was, in the early history of clocks, an enormous leap forward for accuracy over the analog systems such as the flow of water. In such systems balls were counted or weights released from a balance as volumes of water were collected from a 'dripping tap.' The large initial inaccuracies of the crown wheel verge escapement (introduced in the late thirteenth century) were not due to the verge itself, but to the foliot-smoothing mechanism.

Replacements for foliots by the spring balance wheels and, in the mid-seventeenth century, pendulums, left the verge as being the 'weakest link.' Hundreds of inventions or modifications of

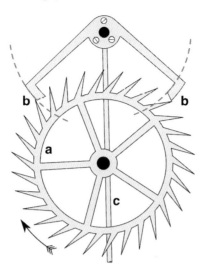

Fig. 2.3 Graham deadbeat escapement. '*a*' is the toothed (escape) wheel driven by the weight or spring; '*c*' is the pendulum (foliot replacement) that is pushed by the wheel and '*b*' the pallets, alternately catching and releasing teeth on the escape wheel, whose radial shape was critical in the deadbeat mechanism for the removal of recoil. (Courtesy of Frederick J. Britten From Wikipedia, 'Escapement' public domain, derivative work, McSush; {PD-US})

escapements were attempted, but only about ten survived to the late seventeenth century. Robert Hooke designed the big advance of the anchor escapement around 1660 (but credit was disputed by William Clement), so called because of its shape. One aspect of the inaccuracy still remaining was the shape of the pallet on the swinging anchor. George Graham made a huge advance in the design of the anchor-shaped escapement, resulting in removal of the wheel backward motion of each 'click' through the design of the pallets or paddles on the anchor (Fig. 2.3). This 'deadbeat' escapement was deemed to be the real beginning of accurate clocks as we know them today.

Although Graham is often credited with inventing the deadbeat mechanism (he likely made a significant improvement to it in 1715), it was actually designed by Richard Towneley and used by Thomas Tompion to build a clock for Sir Jonas Moore. The Towneley mechanism was also used in two regulators made for the Greenwich Observatory in 1676 (when Graham was just three years old) and paid for out of the personal pocket of Moore. Correspondence between Towneley and the Astronomer Royal John Flamsteed included such references.

The cylinder escapement was another brilliant advance that was particularly useful in the confined mechanisms of watches. Again, George Graham is mentioned as the inventor (1726) in some references and Tompion in others. One thread seems to be apparent from researching clock history is that Tompion and Graham, and maybe others of the time, worked so closely together it could be quite difficult to unpick exactly who did and said what when. Knowing how generous in spirit Graham was, he would likely give any advice or involve himself in any discussions over the workbench or a 'coffee break' to impart his knowledge.

As clocks became more and more accurate, it was noticed that a change of temperature, in particular on a pendulum, caused errors due to the expansion or contraction of the metal, which lengthened or shortened the pendulum. George Graham's answer to this was to incorporate a bulb of mercury into the pendulum to replace the customary bob. The amount of mercury could be adjusted to compensate for the variation of length with temperature and therefore maintain a constant center of gravity. A weight-driven clock with a Graham escapement and a mercury pendulum could achieve an accuracy of within a few seconds per day. An alternative compensation mechanism of the bi-metallic gridiron was also variously ascribed in the literature to Graham or Harrison.

Graham had many associations with the Royal Observatory at Greenwich (ROG), London, and its astronomers. In 1884 the ROG became the site of the internationally agreed prime meridian from which all the world's time is taken. James Bradley originally defined the meridian, 0° longitude, as passing through where his telescope stood, and the meridian was adopted by the UK's first Ordnance Survey map in 1801. Sir George Airy redefined the exact meridian position in 1851 to about 6 m to the east of Bradley's, since his instrument was in a room next to Bradley's. At an International Meridian Conference held in Washington, D.C., in 1884 it was decided that this should be the official prime meridian, because the majority of ships were using the Greenwich meridian as their reference point.

The history of Greenwich dates back to antiquity with barrows possibly constructed during the Bronze Age. About 300 coins were also found in 1902 bearing evidence of the Roman Emperors Claudius and Honorius of the fourth century. However, the town of Greenwich was seen as no more than a fishing village with safe anchorage until as late as Henry V (1386–1422, king from 1413 until his death at the age of only 35). On the site of the current National Maritime Museum and Greenwich Park (incorporating the Royal Observatory) the Plantagenet, Tudor and early Stuart monarchs had built a hunting lodge or palace that was variously used for royal purposes, e.g., royal marriages and births, Henry VIII and his mistresses, and as a prisoner-of-war camp and biscuit factory during the English Civil War (1642–1651). The council of Queen Elizabeth I planned the Spanish Armada campaign here. By the time of the Restoration (of Charles II to the throne in 1660) the palace had fallen into disuse and was pulled down.

Under the new king new buildings began to be established and Greenwich Park was redesigned and replanted. Charles II also commissioned the Royal Observatory to be built and the commission went to Sir Christopher Wren (1632–1723). The reason for the building of the observatory was to increase English supremacy of the seas by improving the lot of sailors and merchants, who were desperate to know exactly where they were when at sea, out of sight of land, to avoid catastrophe. The king designated John Flamsteed (1646–1719) as the first Astronomer Royal and the observatory was built to accommodate him and his work. Flamsteed was to "apply himself with the most exact care and diligence to Rectifying the Tables of the Motions of the Heavens and the Places of the Fixed Stars, in order to find out the so much desired Longitude at Sea, for the perfecting of the Art of Navigation."

It is interesting to note here that Sir Jonas Moore (1617–1679), mentioned previously, was an English patron of astronomy, a mathematician and a surveyor known for the huge project of draining the Great Level of the Fens. He used his accumulated wealth and influence to become a patron and was the main driving force behind the establishment of the Royal Observatory of Greenwich (ROG). Moore was disappointed with the productivity of Flamsteed's work at Greenwich and threatened in 1678 to stop Flamsteed's salary if it didn't improve. In fact, Flamsteed was so meticulous that he had still not published his data after forty years!

Astronomers Royal Over the Years

Reverend John Flamsteed, 1675–1719
Professor Edmund Halley, 1720–1742
Dr. James Bradley, 1742–1762
Nathaniel Bliss, 1762–1764
Reverend Nevil Maskelyne. 1765–1811
John Pond, 1811–1835
Sir George Biddell Airy, 1835–1881
Sir William Christie, 1881–1910
Sir Frank Dyson, 1910–1933
Sir Harold Spencer Jones, 1933–1955
Professor Sir Richard van der Riet Woolley, 1956–1971
Professor Sir Martin Ryle, 1972–1982
Professor Sir Francis Graham-Smith, 1982–1990
Professor Sir Arnold Wolfendale, 1991–1995
Martin Rees, Baron Rees of Ludlow, 1995-

At that time, the key to accurate positioning was thought to be knowing the exact mapping of the movement of the Moon and stars, from which could be calculated the longitude. The latitude was easily found from the height of the Sun at midday. Stellar measurements had not been refined since they had been undertaken by Tycho Brahe before the use of telescopes. The use of second-hand building materials and the sale of old gunpowder enabled the project to be financed. Although the money was still short Wren built the beautiful Octagon Room underneath which was the accommodation for Flamsteed.

Calculating and mapping the exact position of the stars in Flamsteed's time was difficult if not impossible to the accuracy required. In addition to knowing the exact time of viewing a star the precise position had to be known. One instrument was the transit telescope, possibly invented by the Danish astronomer Ole Christensen Rømer (1644–1710). In essence, the principle was the same as the meridian circle, transit circle or the quadrant. A round plate or frame inscribed with measured angles, or degrees, was directed at the sky, using a pendulum or other means to relate to the horizontal and aligned along a meridian. The height in degrees of the object (a star) viewed through an attached telescope was noted and related to the time for the determination of the coordinates

FIG. 2.4 Graham mural quadrant at Greenwich. (Courtesy of the National Maritime Museum, Greenwich, UK.)

of the star. The instrument was usually anchored to the ground and clearly difficult to use at sea, although the hand-held sextant with similar principles and split mirrors was standard nautical equipment. It became usable when clocks with an appropriate accuracy were combined with a sturdy anchorage such as a wall, at which point it was known as a mural quadrant. Many models of the transit telescope itself were made and used until recent times. One superb instrument is housed in the ROG.

Not only did Graham develop clocks of great accuracy, he also used his astronomical interests and skills to produce a mural quadrant in 1725 for Halley's early years at Greenwich and later for Bradley (first mounted in the house of his Aunt Pound at Wansted in Essex). The quadrant still exists at Greenwich and is shown in Fig. 2.4. Both Halley's and Bradley's 8-foot mural quadrants are mounted on

Fig. 2.5 Transit telescope probably first started by Robert Hooke and completed by George Graham. (Courtesy of the National Maritime Museum, Greenwich, UK.)

a wall of nine massive stone blocks set into the bedrock of Greenwich hill. Halley, who had made a name for himself studying the less well known stars of the southern celestial hemisphere, originally set his quadrant on the east side of the wall roughly in line with Flamsteed's first Greenwich meridian and facing south.

Halley actually had a difficult start to his early days at Greenwich because the relatives of Flamsteed, mainly his wife, had comprehensively cleared the building of all Flamsteed's instruments, which they viewed as being their personal property. Having been granted a sum of money to re-equip his laboratory, one of Halley's first purchases was a transit telescope, probably first started by Robert Hooke and completed by George Graham. A transit telescope pivots only in one plane, normally north-south, so that as Earth rotates, selected stars become visible in the telescope and the exact time from an astronomical regulator is noted to give the star's position. The telescope consists of a brass tube 5 ft long with an object glass aperture of 1.5 in. (Fig. 2.5).

Fig. 2.6 The Graham clock no. 630 built around 1750 and housed at Greenwich. (Courtesy of the National Maritime Museum, Greenwich, UK.)

James Bradley, the third Astronomer Royal, was granted £1,000 in 1749 raised from the sale of old naval stores to construct a new building and replace instruments that were in poor condition. One of George Graham's clocks was purchased in 1750 for £39 for use by John Bird alongside an 8.5-foot transit telescope. The clock was known as 'Graham 3'. Altogether, four of Graham's clocks were bought for Greenwich. Edmund Halley used the first, bought for £5 in 1721, as a transit clock. The image (Fig. 2.6) is of the Graham clock no. 630 built around 1750 and housed at Greenwich.

Several beautifully ornate Graham and Tompion watches are part of the collection at Greenwich. The cylinder escapement is sometimes used to enable the flat mechanism to be employed. The image in Fig. 2.7 is a signed center-seconds watch with cylinder escapement in a silver pair-case. It is believed to have been made for James Bradley, around 1729, who used it for observations on the aberration of light and the fixed stars. The center-seconds

FIG. 2.7 A George Graham center-seconds watch with cylinder escapement in a silver pair-case. It is believed to have been made for James Bradley around 1729. (Courtesy of the National Maritime Museum, Greenwich, UK.)

hand can be stopped by moving a small lever that protrudes through the side of the inner case. This actuates a spring, which bears on the balance and so stops the watch. The image in Fig. 2.8 shows a pocket watch of Graham's made from brass without dial or hands. It has a full plate fusée movement with a verge escapement. In fact, the discovery of the aberration of light was an accident while Bradley and Samuel Molyneux searched for (and failed to find) annual parallax in the position of the star Gamma Draconis from Molyneux's house on Kew Green, just southwest of London, where Graham's first high-quality zenith sector was placed.

Graham made a bespoke zenith sector for the French astronomer Pierre Louis Maupertuis, 1698–1759 (the request being facilitated by Anders Celsius meeting Graham). Maupertuis ventured close to the north pole in Lapland in 1736 to measure the value of 1° of latitude. The aim was to compare this with that measured near the equator in Peru to determine the ellipticity of Earth. The telescope of the sector was 9 ft tall and hung from the vertex. The observer rested in a reclining position. Maupertuis did indeed confirm Isaac Newton's belief that Earth is an ellipsoid flattened at

Fig. 2.8 A pocket watch of George Graham's made from brass without dial or hands. It has a full plate fusée movement with a verge escapement. (Courtesy of the National Maritime Museum, Greenwich, UK.)

the poles. He has therefore often been referred to as the man who flattened Earth.

Graham was interested in more than just clocks, certainly in astronomy, as mentioned in the previous paragraphs, but also in Earth science, geomagnetism in particular. He made compasses, and the needles he produced were used by many scientists. By making careful observations of a compass needle through a microscope, he discovered the diurnal variations of the terrestrial magnetic field in 1722/23 and their relationship with auroras. That the variations were not just local was confirmed when Celsius in Uppsala, Sweden, simultaneously detected a large magnetic disturbance, a magnetic storm, in April 1741. So internationally noted was Graham for his inventiveness and precision that his works of all kinds are found in collections and museums all over the world and command very high prices whenever they come up at auction. He was elected as a member of the Royal Society in 1720 and chosen to be a member of its council in 1722.

Possibly one of the most important actions of his life for humanity was his financial and professional support for John Harrison. Harrison solved the 'longitude problem,' but not without cost. He devoted his whole working life, against all odds, to the production of a timepiece that would maintain its accuracy at sea. As the commercial, exploring and warring imperatives of many nations increased exponentially from the fifteenth century on, deaths, wrecks and losses of cargo at sea were commonplace. Some such events could be laid at the door of humankind's unfortunate nature to fight and conquer, but many were due to the ships' navigators ignorance of precisely where they were, which fatally resulted in encountering unexpected rocks or no land at all.

Just two measurements, coordinates, are necessary to locate a position. This was known around 2000 B.C. to the Phoenicians and other ancient civilizations who used the position of the Sun, Moon and stars to give the latitude, and a map of stars to find the longitude. Nautical charts created from experience were helpful, but the accuracy was wanting, and most maritime journeys were coastal, with predictable winds and currents or a wide continental shelf to follow until a few thousand years later. Several of the scientific instruments described in the previous chapter were useful in the right hands, especially the sextant, which was invented at the beginning of the eighteenth century.

A system of latitude and longitude for mapping the world was first proposed by Eratosthenes in the third century B.C. Hipparchus was the first to take this on board and to recognize that longitude could be determined by comparing local time with an absolute time. However, despite all efforts, for almost 2000 years no one came up with a clock sufficiently accurate and usable at sea. Ever increasingly accurate star maps were pursued in an attempt to solve 'the intractable problem.' As already noted, Flamsteed was specifically given the task of doing just that, since King Charles II had been keen to investigate a suggestion by a Frenchman, the Sieur de St. Pierre, that a knowledge of the position of the Moon against the backdrop of stars would give the longitude. This was also supported by Robert Hooke, the polymath, who encouraged Flamsteed. It was also being accepted throughout the scientific and maritime community that it would never be possible to produce a marine chronometer, but a great prize was being offered by several nations' rulers for its solution. Then came the disaster! It was one of the worst in British maritime history and was due to the inability of the navigators to know their precise position.

Fig. 2.9 The commemoration stone for Admiral Shovell's on Scilly. (Courtesy of Jeff Harries.)

During the war of the Spanish Accession a combined naval force of British, Austrian and Dutch ships were dispatched to Toulon, but in spite of some success, it did not go well, and the British fleet of twenty-one ships, led by its commander-in-chief with the unforgettable name of Sir Cloudesley Shovell was ordered back to Portsmouth. Toward the latter part of the journey the weather worsened, the ships were blown off course and their location was uncertain. Legend has it that one sailor recognized where they were and let it be known that he thought the navigator was wrong. He was hanged for inciting mutiny! Four of the ships, including "The Association," on which sailed Admiral Shovell, hit rocks off the Isles of Scilly and all 800 men on board drowned. The commemoration stone for the place of Shovell's demise on Scilly is shown in Fig. 2.9. Possibly as many as 2,000 souls in total were lost, and bodies were washed ashore for days afterwards.

What a wakeup call this was for the British navy and the government! An incentive was needed to concentrate the best minds of the day, and what better than money? By act of Parliament the longitude prize was offered in three main parts: £ 10,000, £ 15,000 and £ 20,000 for longitude to an accuracy of within 60, 40 and 30 nautical miles, and additional money up to £ 2,000 toward promising lines of research. The Board of Longitude was set up to judge submissions. Many still clung to the precise mapping of the stars and movement of the Sun and Moon as the only solution, believing that the quest for a suitable seagoing chronometer was doomed to failure. Unfortunately, among the serious doubters was the Astronomer Royal, Nevil Maskelyne, who was appointed to the board and proved to be a major hindrance to progress toward a solution.

Such a large purse was bound to attract all sorts of suggestions and applications, including crackpot ideas. That was not all bad, though, as it raised awareness of the project even if it did cause some embarrassment for those seeking a serious solution. Among the most quoted 'sillies' was one that recommended anchoring a huge number of barges across the ocean from which rockets could be fired at 'known' intervals of time (actually, an early suggestion that probably led to creation of the longitude prize) and a 'Powder of Sympathy' that operated over any distance and included carrying a wounded dog on board that would howl when the dog's bandages, left onshore, were plunged into the powder. The latter was, apparently, just an anonymous joke. Many suggestions just wasted time, such as including designs for perpetual-motion machines or improving ships' rudders or sails. By the time a young hopeful arrived in London in 1730 aspiring to present his ideas to the board, the group had given up meeting and was just sending out polite letters of rejection.

The young hopeful was John Harrison. He was born in 1693 in Yorkshire and lived with his parents on the estate, Nostell Priory, of a rich landowner who employed Harrison senior as a carpenter. Little wonder, then, that Harrison junior became a skilled worker in wood. He was also musical, played the viol, became a choirmaster and tuned church bells. He enjoyed learning and was encouraged to do so. Something in that learning sparked his obsession with clocks, and, using his knowledge of carpentry and the properties of wood, he made his first almost entirely of that material, in which even the teeth were durable.

Harrison had his share of personal bad luck when his first wife Elizabeth died and their son John followed her later at only eighteen years old. However, Harrison married another Elizabeth just six months after his first wife died, and they produced two children, John, who later gave great support to his father, and Elizabeth, about whom little is known. John (senior) worked closely with his brother James to produce some fine clocks, each with their new inventions to further develop the accuracy of timekeeping. John began thinking about the longitude problem, and when he was happy with his ideas of springs and counterbalances, he set off to London.

As a last resort to gain access to the board he sought out one member and knew just where to find him—in the Royal Observatory at Greenwich. Halley was a congenial and generous man and generally open-minded. He listened to Harrison's design for a marine clock with great interest. But he knew it would be difficult to get the board to even vaguely consider a mechanical solution when most were hell-bent on using the stars.

As Harrison had never been great with words, Halley sent him to Honest George Graham with instructions on how to approach one of the most revered clockmakers of the age, who was kindly and would understand the proposal. At the meeting Harrison feared that Graham would steal his ideas and Graham feared having to deal with another crackpot. In the history of clocks and the lives of sailors the day was most significant. Once they broke the ice, both horologists began to trust each other and discussed and studied the inventive designs before them. The mighty George Graham and the young carpenter then had dinner together, finally resulting in Graham giving great encouragement and a loan from his cash box, without interest or time limit. Without this most productive of meetings and the generosity of Honest George the world might have had to wait for many more decades to see an improvement in the safety of mariners.

After about five years, Harrison produced H1, a marvel to behold in mechanical science and appearance. It looked like no other before it with its shiny brass fittings and contra movements of springs and levers (Fig. 2.10). In spite of the resistance of the board, mostly from Maskelyne, Harrison persisted with design improvements and the production of glorious clocks (H2, 3, 4 and 5). This eventually led in 1772—with the assistance of his son, Parliament and King George III ("by God Harrison I will see you righted!")—to the award (in parts over many years) of the Longitude Prize money.

Harrison died in 1776 at age 83. Larcum Kendall (1721–1795) produced further watches, K1, 2 and 3 as part of the Harrison legacy, but John Arnold (1736–1799) was responsible for designing an

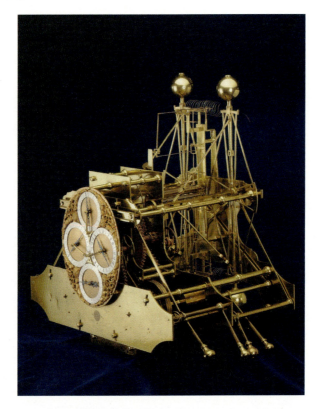

Fig. 2.10 John Harrison's chronometer, H1. (Courtesy of the National Maritime Museum, Greenwich, UK.)

acceptably accurate and practical watch for general use. H1 is now safely housed and displayed in the Royal Observatory at Greenwich and is priceless. Even a modern replica was sold at auction to the director of Charles Frodsham & Co., London, for $ 904,000.

George Graham made another, albeit indirect, improvement to the marine chronometer. He took on an apprentice, 14-year-old Thomas Mudge (1715–1794), who was clearly so well trained by Graham that he became a freeman in 1738. In 1750 he moved to his own premises in Fleet Street where Matthew Dutton joined him as a partner four years later. He was another of Graham's apprentices. Mudge's new lever escapement mechanism was a major advance and he retired from the business in 1771 to work on a marine chronometer, for which he was granted £500 to develop. In 1779 he produced two such instruments.

Once again, Maskelyne attempted to block the progress and deemed the chronometers unsatisfactory. However, in 1776

Mudge was made clockmaker to the King George III. Mudge was nearly blind when he made the last two of his famed three marine timekeepers in the late 1770s. The first one, known as "Mudge No. 1," is in the collection of the British Museum. The second, "Mudge Blue," is in the Mathematisch-Physikalische Salon in Dresden, and there is a third one, "Mudge Green," which was completed in 1779 and was sold for $ 1,240,000 in 2004. All three are celebrated in the history of the pursuit of the elusive Longitude Prize.

"Blue" and "Green" are identical except for the corresponding colors of their shagreen (tough leather) cases. They covered their cases with different colors so that they could tell them apart. The twinning enabled Mudge to determine if variations in the timekeepers' rates were caused externally or internally. Making two identical timekeepers was also a requirement of the board that was in charge of issuing the major prize, which they never did in Mudge's lifetime. Only after Mudge's son championed his father's work was a kind of consolation prize awarded.

Thomas Mudge's portrait in his later years is shown in Fig. 2.11. Two years before his death a committee of the House of Commons overruled the board and awarded £ 2,500, thus ensuring the Mudge clocks to be part of the story of chronometers available to mariners. Mudge 1 is shown in Fig. 2.12. The British Museum's description includes the following:

> In a letter to his patron, Count Maurice von Brühl, dated 23 August 1774, Mudge wrote, "I acknowledge it [the case] would have been better of brass. My only reason for making it of wood, was to save money, of which I have had, at no time, much to spare." The machine was tested in private trials at Greenwich between June 1774 and February 1778 but failed because the mainsprings kept breaking. On 1 March 1777 Nevil Maskelyne, the Astronomer Royal, stated that it had gained 1 min and 19 s while on trial at Greenwich for 109 days—an average of less than one second per day. He said that the machine was "greatly Superior in point of accuracy to any timekeeper which hath come under my inspection." As a result of its performance Mudge was awarded a payment of £500 by the Board of Longitude. Following a report by a House of Commons Select Committee in 1793, Mudge was awarded a further £ 3,000, but his death the following year precluded any further work. It was not to be the end of the story, though: using the money, Thomas Mudge junior set up a small manufactory where some twenty-seven copies of his father's chronometer were made.

FIG. 2.11 Thomas Mudge's portrait in his later years. (Courtesy of the Trustees of the British Museum.)

FIG. 2.12 Mudge clock number 1, 'Mudge 1'. (Credit: © Trustees of the British Museum.)

FIG. 2.13 Marine chronometer no. 2741 by Breguet et Fils. (Courtesy of the Trustees of the British Museum.)

The legacy of the pioneering clockmakers continued. Figure 2.13 shows a chronometer made by the famous Frenchman, Breguet. The British Museum writes:

> Towards the end of the eighteenth century improvements in the timekeepers used for finding longitude at sea were, to a large extent, achieved by English chronometer makers, particularly John Arnold and Thomas Earnshaw. Nevertheless, the great French clockmakers were also striving for the same goal, and makers such as Ferdinand Berthoud, Pierre Le Roy and Henri Motel all played their part. One of the most celebrated of French clock and watch makers was Abraham-Louis Breguet, and although his workshops were primarily concerned with the making of watches, he also made a number of marine chronometers.
> This example was completed in 1813 and used as an experimental piece by Breguet himself until 1822. It was presented to his friend Monseigneur Belmas, Bishop of Cambrai, to whom it was inscribed.

74   Orrery

FIG. 2.14 John Arnold's chronometer with eight-day movement and a spring detent escapement. (Courtesy of the Trustees of the British Museum.)

Two other names strongly associated with the development of usable chronometers are John Arnold and Thomas Earnshaw.

John Arnold was a watchmaker and principle chronometer maker of his day. His business was in Devereux Court, The Strand, thus adding to the concentration of clock-making history in the nearby Fleet Street area. He made a repeating watch set in a ring for King George III. Figure 2.14 shows one of his chronometers with eight-day movement and a spring detent escapement. The beautifully crafted mahogany case is typical of the care taken by craftsmen of the time.

Thomas Earnshaw (1749–1829) was Arnold's rival in an argument as to who invented the spring-detent escapement. The British Museum writes of Earnshaw:

> Born in Ashton-under-Lyne [Greater Manchester, England] in 1749, Earnshaw served an apprenticeship in London, which he completed in 1770; he then worked as a trade watch-finisher and escapement maker of extraordinary ability. In contrast to John Arnold, who had

Fig. 2.15 Thomas Earnshaw's chronometer with a one-day spring-driven movement and Earnshaw split-bimetallic temperature balance. (Courtesy of the Trustees of the British Museum.)

a somewhat privileged early career, Earnshaw came up the hard way. He married before he had finished his apprenticeship, and within four years had three sons. By 1774 he was facing the debtors' prison and fled to Ireland, but in the end his conscience overcame him and he returned to the Fleet Prison in December 1774.

However, Earnshaw survived all this to create a thriving and successful business in 119 High Holborn, well away from Arnold. One of his chronometers is shown in Fig. 2.15, a one-day spring-driven movement with Earnshaw's split-bimetallic temperature balance.

All of the clockmakers described had their own masterful influence on clocks, watches and chronometers right up to the twentieth century.

Graham the Scientist and Astronomer

Should we say George Graham the clockmaker or George Graham the astronomer? Some say that his contribution to astronomy was of even greater importance. The construction of sectors and quadrants, now taking pride of place in prestigious museums and observatories, and the company he kept, Molyneux, Hooke and Astronomers Royal, already give hints. Indeed he was one of the few clockmakers to reach the dizzying heights of the Fellowship of the Royal Society, which boasted many of the top names in astronomy as members.

Not only did Graham support astronomers through the production of instruments, he also made detailed observations and produced such works as "An Occultation of Alderbaran by the Moon." This paper, from 1722–1725, was part of his project to test the suitability of his newly invented temperature-compensating mercury pendulum. Even Maskelyne, so antagonistic toward Harrison and his marine chronometer, praised the device as "an excellent astronomical clock with a gridiron pendulum made by Mr. John Shelton under the direction of George Graham ... by which transits of the heavenly bodies over the meridian are observed." By altering the length of the pendulum, Graham bridged the gap between ordinary clocks based on solar mean time and sidereal time (calculated from the apparent movement of the stars) that was of essential use to astronomers. The *Gentleman's Magazine* for 1751 described the best such instruments in France, Spain, Italy and the West Indies as copies by English artists of Graham's originals. Toward the end of his life Graham devoted almost all of his time to astronomy.

Reading histories and descriptions of Graham's life and works is fascinating, but to get a feeling of the times, one can become a little closer by perusing the original Royal Society's Philosophical Transactions of the time through the old English writing, grammar and sentence construction. Here are titles of a few transcripts of submissions to the Royal Society relevant to Graham, although in modern type:

1. "An account of some observations made in London by Mr. George Graham, F.R.S. and at Black-River in Jamaica, by Colin Campbell,

Esq; F. R. S. concerning the going of a clock in order to determine the difference between the lengths of isochronal pendulums in those places. Communicated by J. Bradley, M. A. Astr, Prof. Savill. Oxon. F.R.S."
2. "The same eclipse observed in Fleetstreet, London. By Mr. George Graham, F.R.S.Phil."
3. "Observation of an extraordinary height of the barometer, December 21, 1721. By Mr. George Graham, Watchmaker, F.R.S."
4. "Observations of the dipping needle, made at London, in the beginning of the year 1723. By Mr. George Graham, Watchmaker, F.R.S."
5. "An account of some Magnetical Observations made in the months of May, June and July, 1723, in the Atlantick or Western Ocean; As also the description of a water spout, By Mr. Joseph Harris. Communicated by Mr. George Graham, F.R.S."
6. "An occultation of Jupiter and his satellites by the moon, October 28, 1740, in the morning; observed at Mr. George Graham's, F.R.S. house in Fleetstreet, London, by Dr. Bevis and Mr. James Short, F.R.S."

Clearly, what was possible then would not be so easy now with the current levels of light pollution.

The Proto-Orreries

Models or depictions of the Solar System have been attempted since antiquity, even though early examples involved erroneous or extremely limited contents. The physician, antiquary, archeologist and friend of Sir Isaac Newton, William Stukeley (1687–1765), reported that Richard Cumberland (1631–1718) had made a planetarium (not in the current sense of an astronomical theater). He was rector of All-Saints, Stamford, and later bishop of Peterborough. Some references suggest that his grandchildren played with the model until it dropped to pieces and was lost. Stukeley also drew a diagram of a geared model of the Sun/Earth/Moon system mounted inside a circular tablestand that was made by Dr. Hale, a fellow of Corpus Christi College, Cambridge, in 1705.

Stukeley claimed that Dr. Hale was the true inventor of the 'orrery.' There was some bitterness and rivalry as to who could claim to be the first constructor of the model. In an account of the life of Stephen Hales, D.D., F.R.S., it was mentioned that "Mr. Hales was equally assiduous and successful in the study of Astronomy, for, having acquired a perfect knowledge of the Newtonian system, he contrived a machine to demonstrate it, which was constructed of brass, and moved by wheels, so as to represent the motions of all the planets, upon the same principles, and nearly in the same manner as the machine afterwards constructed by Mr. Rowley, master of Mathematics to King George I, which was absurdly called an Orrery, because an Earl of Orrery was Rowley's patron."

As more astronomical awareness developed, more complex displays became possible. These became known as planetaria (planetariums) or telluriums (tellurions, tellurians), some displaying the whole of the sky as seen. As clockmakers led the way in constructing beautiful and accurate timepieces, one maker in particular created the first geared model (but see the Antikythera mechanism previously described and Dr. Hale, above) that became known as an orrery or proto-orrery. Although it was a most intricate and ornate instrument, it displayed just the Earth, with its orbiting Moon, going around the Sun. The mechanism was made by the recognized genius of his time, George Graham. Thomas Tompion had some influence as they worked very closely together in their clock-making business.

Had it not eventually been blessed with the prestigious name of 'orrery,' the device might never have made such a splash in history, although it additionally became popular for educational and demonstration purposes due to the underlying principle of the universal theory of gravity published by Newton in 1687. A brief delving into its history often throws up firstly that the model was due to the fourth Earl of Orrery, thereby implying that he made it. Eustace Budgell, a writer of the day, is quoted as follows on the subject: "The Instrument which was invented by him, and bears his name, is an undeniable Proof of his [the Earl of Orrery] mechanick Genius." Others say that the model was built at the request and patronage of the Earl of Orrery, both possibilities leading to the label orrery. Graham's geared model was built around 1710 (1704 or 1703–1709 in some references) and was given to the eminent instrument maker John Rowley of London to send to Prince Eugene of Savoy along with other Rowley instruments. Rowley was commissioned to make a copy for his patron, Charles Boyle, fourth Earl of Orrery, and the completed instrument was presented to Charles' son John. These stories illustrate the confusion as to what really happened, and even now there is uncertainty in

Fig. 2.16 George Graham proto-orrery, 'G' model. (Courtesy of the Adler Planetarium & Astronomy Museum, Chicago, Illinois.)

Fig. 2.17 The second proto-orrery signed by George Graham and Thomas Tompion (G & T model). (Courtesy of the Museum of the History of Science, Oxford.)

the literature concerning some details. It is known that Graham produced two geared models, one just signed Graham (let's say the G model; Fig. 2.16) and another signed Graham and Tompion, (G & T model; Fig. 2.17). It is likely that Charles Boyle, fourth Earl of Orrery, was a patron of George Graham.

These first models, G & T and G, could not be called true orreries, since the geared models had not been so named until after their production. They cannot, strictly, be called planetaria either, since they do not have planets other than Earth. Therefore they have been referred to as proto-orreries.

The G & T model is now in the Museum of the History of Science, Oxford. It was probably signed Tompion as well as Graham, because all objects for sale made on the premises had to have Tompion included in the signature.

The clockwork-driven G model is now displayed in the Adler Planetarium and Astronomical Museum in Chicago in the United States. As it is not as ornate and has some flaws, it was probably a prototype of the G & T, hence only signed and kept by Graham. Although made by clockmakers, the motion of orreries was often driven by a handle, not a clockwork mechanism.

The G & T model was constructed for the Julian and Gregorian calendar, while the G model was only for the Julian, although the latter was, in 1752, converted to the Gregorian, probably for ease of trade with Europe. Prince Eugene and John Rowley were involved.

More will be said in the next chapter about the Earl of Orrery, but who were Eugene and Rowley?

Prince Eugene

Whatever else is uncertain, it seems a certainty that Prince Eugene, a prominent and revered warrior and statesman, played his part in the procurement, ownership and care of the G & T proto-orrery. But why should he be the recipient of one of the first prestigious models? Why was he so worshipped?

To give him his full title, Prince Eugene Francis of Savoy-Carignan-Soissons was born in Paris to Italian parents in 1663. He was the youngest son of Olympia Mancini, the niece of Cardinal Mazarin, and Eugene Maurice of the Savoy-Carignan-Soissons. He had four elder brothers and two younger sisters. His mother was favored at the court of Louis XIV, where Eugene was brought up, but she fled to Brussels in 1680 when she was accused of being a poisoner. For the next three years he was under the care of his grandmother, Marie de Bourbon, who threw him out of the house for refusing to become a priest, after which he fled Paris (disguised as a girl), as he was refused entry into the army of King Louis XIV

Fig. 2.18 Prince Eugene Francis of Savoy-Carignan-Soissons. (Jacob van Schuppen, 1718.)

on the grounds of a weak constitution. Leopold, of the Habsburg monarchy, allowed him to join the imperial Austrian army as a Volontar in 1663, where his cousins protected and financed him, as did the head of the house of Savoy, Duke Victor Amadeus II. At last he was a soldier and not a priest as his parents had intended; indeed, he was known as the Abbe de Savoye in 1678.

He first gained fame in Europe after his victory against the Ottomans at the battle of Zenta in 1697. Prince Eugene is shown in the stunning portrait (Fig. 2.18) by the painter Jacob van Schuppen. Eugene was best known for his courage and success in battles such as against the Turks laying siege to Vienna, campaigning in Hungary and even against the French in the Spanish War of Accession, when he joined forces with the First Duke of Marlborough (John Churchill) at Blenheim (1704), the main confrontation of the war. A celebration of the victory at Blenheim is shown in Fig. 2.19 and is described in the attached legend at the British Museum:

FIG. 2.19 A 1704 print of the celebration of the victory at Blenheim with Prince Eugene and John Churchill, Duke of Marlborough. (Courtesy of the Trustees of the British Museum.)

Responding to appeals from Vienna, which was threatened by French and Bavarian forces, the English commander, John Churchill, Duke of Marlborough, marched his army from the Netherlands to Bavaria and joined forces with the Austrian general, Prince Eugene of Savoy. At Blenheim their combined army overwhelmed a Franco-Bavarian

FIG. 2.20 A 1709 etching of Prince Eugene and the Duke of Marlborough once more fighting together at the battle of Tournay. (Courtesy of the Trustees of the British Museum.)

force under Marshall Tallard and the Elector of Bavaria. For the first time in two generations the French suffered a crushing defeat, and the results were immediate and far-reaching. Bavaria was conquered and Vienna saved. The territorial ambitions of Louis XIV beyond the Rhine were checked, and France was placed on the defensive.

More than once Eugene, using his recognized skills in war, joined with the duke in battle, which clearly resulted in their close friendship. Figure 2.20 is a print of a 1709 etching of the Duke of Marlborough and Prince Eugene of Savoy on horseback, foreground left, surveying the bombardment of Tournay in 1709. Figure 2.21 shows a portrait of John Churchill, Duke of Marlborough, in 1704.

The last engagement fought by the two commanders together was the battle at Malplaquet in France (1709). Probably harking back to his resentment of King Louis XIV, Eugene in his capacity as general of the Austrian cavalry led his army to remove the French from northern Italy. This belonged to Spain at the time. Acknowledged as one of the greatest war strategists and generals of his time, he spent the last years of his life on his great project of reforming the Austrian army and lived in luxury in the Great Palace of the Belvedere in Vienna.

84   Orrery

Fig. 2.21 John Churchill, Duke of Marlborough, in 1704. (Adrian van der Werff.)

When the Treaty of Rastatt (1714) ended hostilities between France and Austria, Eugene mixed with scholarly men (such as Gottfried Leibniz, Jean-Baptiste Rousseau and Montesquieu), who were always keen to obtain patronage from the now wealthy prince. His art collections comprised sixteenth and seventeenth century Italian, Dutch and Flemish masters, and his library contained over 15,000 books, 237 manuscripts—he was particularly interested in natural history and geography—and a huge collection of prints. In 1726 he bought Schloss Hof, a beautiful tract of land near the Austrian/Slovakian border, and hugely extended and developed it into a stunning summer palace and hunting lodge. Prince Eugene's baroque palace gardens with their vast terraces, ornamental shrubbery, lovely fountains and fine sculpture were known as the most beautiful gardens of the Danube monarchs by 1730. When he died in his city palace in Vienna in 1736, probably from pneumonia, most of his accumulated collections went to his niece, Princess Maria Anna Victoria. She immediately sold everything since she had no interest in, or respect for, her uncle and his exploits.

John Rowley

It is easy when scanning written works on astronomy to receive the impression that the only thing Rowley ever did was to be associated with the first geared models of orreries. Nothing could be further from the truth. The following information is a combination of facts gleaned from several sources.

John Rowley (1668–1728) was born in Lichfield, Staffordshire, and had three older brothers and a sister. His father, William Rowley, was a sword cutler (a maker of sword blades). At the age of 14 he became apprentice to the London mathematical and optical instrument maker Joseph Hone in the Broderers Company (or maybe to the instrument-maker W. Winch). He became a freeman of the Broderers in 1691 and began his own business in 1704 (or possibly 1698) in his shop, The Sign of the Globe, in Fleet Street. His patron was Prince George of Denmark, the husband of Queen Anne of Great Britain, from 1704 until the Prince's death in 1708.

Rowley produced a wide range of mathematical, drawing, surveying and gunnery instruments to be given to Christ's Hospital mathematical school and other groups. He made a sextant and other instruments for the observatory at Trinity College, Cambridge, which led to him being recommended for membership of the Royal Society. He was then asked to report on the instruments of John Flamsteed during 1711–1713. He made large sundials for Blenheim Palace and St. Paul's Cathedral. He became the principle supplier to one of the finest collections of instruments amassed by the Earl of Orrery, who probably commissioned Rowley to produce a tellurian of the type built by Graham, but larger.

Again for the Earl, he created (Fig. 2.22) a most pleasing Butterfield Dial for latitudes 40–55° north. The central section is engraved with the arms of the house of Orrery, which includes lions and the family motto: HONOR VIRTUTIS PREMIUM, honor is the reward of virtue. The compass has cardinal points and the north point is offset 10° west of north to allow for magnetic variation, among other attributes. It is signed J. Rowley LONDINI. In fact, he

Fig. 2.22 The Butterfield Dial for latitudes 40–55° north created for the fourth Earl of Orrery by John Rowley. (Courtesy of the National Maritime Museum, Greenwich, UK.)

made a second almost identical dial for the Duke of Marlborough, which has a black leather case and his coat of arms on the dial. The place names on the back are connected to Marlborough's campaign in the Low Countries during the War of the Spanish Succession.

Rowley also produced magnificent orreries for the Habsburg emperor Charles VI, for Peter the Great, and for the East India Company. For the price of £1,000 he created an elaborate model in 1722 called the great solar system, which included Mercury and Venus. The orrery ebony case with silver ornaments was sold for £500. It is likely that this model never actually reached India and never even left England due to the death of an intermediary in the purchase. Rowley left his shop, which he gave to Thomas Wright, so that he could concentrate on his duties as 'Master of the Mechanicks' to George I, in particular, for the repair, maintenance and building of water engines (water pumps). While collaborating with the Board of Ordinance he made a great contribution to standardizing the measure of the English yard and standardizing calibration of land and sea cannon. One scientifically worrying aspect was that during a scholarly meeting, Samuel Johnson praised Rowley for his perpetual-motion experiments!

On the Trail

It does appear that Prince Eugene visited England, possibly to assist his friend the Duke of Marlborough in recovering his reputation and standing in court, which failed. Then he went to London on a scientific instrument and art collecting spending spree. He popped into both the Tompion/Graham establishment (the Dial and Three Crowns) as well as into the nearby Rowley shop (Sign of the Globe). The purchased G & T model was to be shipped by Rowley to Austria, along with some of Rowley's other instruments. While the model was with Rowley, he was commissioned by the Earl of Orrery to make a copy for him, and Rowley then named the model an orrery after his patron.

It had been suggested that Sir Richard Steele (Irish essayist, 1672–1729) came across Rowley's model in a presentation delivered by Rowley and, knowing nothing of the Graham model, named it an orrery in honor of the Earl of Orrery to popularize it. In his newspaper, one of the precursors of *The Statesman,* Steele wrote: "All persons, never so remotely employed from a learned way, might come into the Interests of Knowledge, and taste the Pleasure of it by this intelligible Method [the orrery]. This one Consideration should incite any numerous Family of Distinction to have an Orrery as necessarily as they would have a clock. This one engine would open a new Scene to their Imaginations; and a whole Train of useful Inferences concerning the Weather and the Seasons, which are now from Stupidity the Subjects of Discourse, would raise a pleasing, an obvious, an useful, and an elegant conversation."

However, Steele even referred to Rowley as 'the Honest Man,' which might also indicate some confusion with 'Honest' George Graham.

These facts were rectified by the lecturer and writer Desaguliers (1683–1744), who was also a scientist and inventor of high repute, although he continued to attribute the actual naming of the orrery to Steele when it was, quite possibly, Rowley (but see later). Desaguliers stated in "A Course in Experimental Philosophy, 1734" that "Mr. George Graham (if I am rightly informed) was the first person in England, who made a movement to [show] the Motion of the Moon round the Earth, and of the Earth and Moon round the Sun, about 25 or 30 years ago. In this machine everything was well and properly executed." He continued to detail its features; to show the phenomena of day and night and their gradual increase and decrease, the seasons, the place on Earth where the Sun is successively vertical and so on.

Desaguliers was generous over Steele's lack of awareness of Graham, "Sir Richard Steele, who knew nothing of Mr. Graham's Machine, in one of his Lucubrations, thinking to do justice to the first Encourager, as well as the inventor, of such a curious instrument, call'd it an Orrery, and gave Mr. J. Rowley the Praise due to Mr. Graham." A suggestion had also been made that Queen Anne commissioned it as a presentation to Marlborough. Then, either Marlborough gave it to Eugene, or Queen Anne changed her mind and gave it to Eugene instead of Marlborough, who was falling out of grace with the court. However, historical circumstances make the involvement of Queen Anne difficult to believe.

There are several references that add weight to the fact that the G & T model was commissioned by Prince Eugene. What is certain is that the G & T model arrived in Eugene's stunning Stadtpalais in Vienna or Belvedere Palace in the city's outskirts to join the rest of the vast collection sometime between 1705 and 1712, depending on the truths of the stories of gifts or purchases. Figure 2.23 shows the palace in 1753.

Prince Eugene was unmarried. Upon his death in 1736, since he had not made a will that might have benefitted those who had served him, his huge fortune, buildings and collections passed to Princess Anne Victoria of Savoy, the daughter of his oldest brother. As mentioned earlier, she had felt no fondness for her uncle and promptly sold the lot, sometimes for knock-down prices. No records of the sales were made, making tracking of the provenance of some of the artifacts difficult or impossible. However, it seems certain that the G & T model was acquired from Eugene's estate by Emperor Charles VI, as a 'Copernican Planetary System' was included in a record of royal possessions obtained from the

Honest George, Chronometers and the Mystery ...

Fig. 2.23 Belvedere Palace in 1753

estate of Prince Eugene. The listing of emperors below might assist in following the trail. Note that much confusion can occur, since Charles and Karl, and Francis and Franz, are same names, and where the numbers for the same monarch differ it is because the 'countries' that they represented changed during the constant round of wars and takeovers.

Holy Roman Emperors from the House of Habsburg from Ferdinand II to Karl I

Ferdinand II	1619–1637
Ferdinand III	1637–1657
Leopold XII	1665–1705 (also known as Leopold I)
Joseph I	1705–1711
Karl (Charles) VI	1711–1740 (also known as Karl [Charles] III)
Maria Theresia	1740–1780 ("Habsburg-Lothringen" or "Hapsberg-Lorraine")
Joseph II	1765–1790
Leopold II	1790–1792 (also known as Leopold XIII)
Franz (Francis) II	1792–1806 (resigned and started his own empire)

Habsburg Emperors of Austria

Franz (Francis) I	1806–1835 (same person as Franz II)
Ferdinand I	1835–1848
Franz Joseph I	1848–1916
Karl I	1916–1918

The next likely appearance of the G & T was in the possession of Stephan, Grand Duke Francis III of Tuscany, husband of Maria Theresia (Theresa). When he became Emperor Francis (Franz) I Stephan asked Johann Georg Nesstfelt, a master carpenter and cabinet maker, to repair an English planetary model whose mechanism had been spoiled. Nesstfelt completed the repair in 1753, and the model was placed in the exhibition of the Imperial Library. There was some postulation that it was a Rowley model, but most likely it was the G & T. After the emperor's death the royal collections were divided up and redistributed by his wife. In 1765 she ordered that the Kunstkabinett collection must be relocated in several salons next to her newly established Augustinergang in the Hofburg. The collections were displayed in nine rooms. In 1791 the collection was moved from the Hofburg to a hall of the Schweizerhof. Over the next several years the collections were brought together in the left wing of the Hofbibliothekgebäude. In 1806 the collections were divided once more and the Mechanisch-Physikalische Kabinett was transferred to the Physikalisch-Astronomische Kabinett. In 1810 it appeared in the Schweizerhof under the astronomical tower in the Imperial City until the Kabinett was disbanded in 1886.

The G & T model could have been moved at any time from where it had been on display in the Imperial Library. The English instrument vanished from the collections sometime during the moves, a guess being that it was 'nicked' and sold by a member of the staff. It is no wonder that from the above complex sequence of events, an instrument in the collections would not be accompanied by precise provenance; and it is all just a best guess anyway. Then there was historical silence with regard to the G & T proto-orrery until around 1930, when it was purchased by a dealer, Jesse Myer Botibol. Botibol, possibly of Austrian extraction, resided at various locations in London and sold artifacts at Christie's auction sales.

Botibol had purchased the model from a monastery that had owned it from a date that coincided closely with the date of its disappearance from the Imperial Library. It was not unusual for monasteries to collect astronomical instruments. The monastery was the Benedictine monastery Erzabtei St. Peter in Salzburg. Ownership by the Salzburg monastery was confirmed in correspondence from Pater Berthold Egelseder of the nearby Benedictine monastery in Michaelbeueren. In a personal communication during the preparation of this work Dr. Hirtner of the Archiv der Erzabtei St. Peter generously scanned the diaries and account books from 1886 to 1930, but no relevant record revealed itself.

Several dealers were involved in attempts to acquire the model, but it eventually appeared with a dealer, Francis Harper, residing at Wickin's Manor, Charing, Kent, England, who had purchased it from either Botibol or the London dealer Solomon Nathan Nyburg. Much prevarication continued between dealers and, on the advice of Sir Geoffrey Callender, first director of the National Maritime Museum, the curator of the Museum of the History of Science, Oxford, Frank Sherwood Taylor was to visit Harper in 1943. Unfortunately, Harper died unexpectedly and his wife consigned the instrument to be sold at Sotherby's auction sales in 1948, Lot 174. The Oxford Museum bought the model using funds and private donations. It is now on permanent display there.

The church of St. Peter within the monastery of Erzabtei in Salzburg was founded by St. Rupert, a Franconian missionary, around A. D. 700 and may have been built upon, or expanded, an existing local religious faction created a few centuries before. It is the oldest community of monks on the Austro-Germanic soil. The buildings were entirely destroyed by fire in 1127 but resurrected between 1130 and 1143 by Abbot Balderich. Many modifications were made over the centuries to produce the stunning church it is today (Fig. 2.24), housing the oldest library of Austria as well as many musical and other collections. Its ancient cemetery is famous, and among those resting here are Maria Anna, or Nannerl, Mozart (Mozart's sister) and Michael Haydn (composer Joseph Haydn's younger brother).

Michaelbeuern Abbey dates back to at least 736, when it was apparently damaged during the Hungarian wars. A fire caused much destruction in 1346, but the abbey saw much better days from the seventeenth century on, and the magnificent baroque

FIG. 2.24 The magnificent church of St. Peter within the Erzabtei in Salzburg. (Courtesy of Creative Commons Attribution-Share Alike 2.5 Generic license. Attribution, Andrew Bossi.)

high altar (Fig. 2.25) was built in 1691. Today it is a thriving Benedictine abbey and owns many businesses such as a farming complex, heating device project and a part share in a brewery. It is the cultural and economic center of Dorfbeuern.

The labeled outline map of Austria (Fig. 2.26) indicates the known places associated with the G & T proto-orrery.

The G model had a much simpler and known history. It was likely that it passed to Graham's benefactors following his death in 1751, but nothing more was heard of it until it appeared, then promptly disappeared, some years before the G & T model, only to appeared again in 1915 in a purchase by the dealer Simmons from the Stevens Galleries in London. It was then sold again at auction in New York, Lot 1288, at a sale of The American Art Association of the properties of Henry Symonds for $825 to an agent, Otto Bernet. 'Honest George Graham's Famous Orrery

Honest George, Chronometers and the Mystery ... 93

FIG. 2.25 The magnificent baroque high alter in Michaelbeuern Abbey, Austria, built in 1691. (Courtesy of Creative Commons Attribution-Share Alike 3.0 Unported license. Attribution; Werner 100359.)

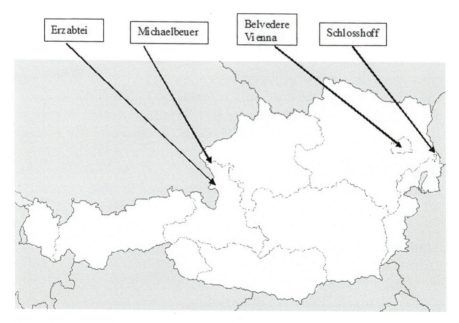

FIG. 2.26 An outline map of Austria indicating the known places associated with the G & T proto-orrery. (Illustration by the author.)

Clock' again disappeared until 1941, when it was offered for sale to the Adler Planetarium and Astronomical Museum in Chicago. The offer was made by an importer of antiques, Samuel Wilson, on behalf of the owner, Byron de Forest, an engineer in the Line Design Section of the Commonwealth Edison Company. The offer was not taken up at that time and the instrument was stored in the company's vault. Five years later the G model was offered once more by de Forest to the Adler museum's F. Wagner Schlesinger, accompanied by a fantastical story of provenance by de Forest that was totally untrue, involving complete tracking of the model through families and heroes. When the price had been reduced, the Adler museum bought it, and it is there to this day on permanent display.

A cursory glance at the history of the orrery and the trail of the G & T model could easily throw up a smokescreen as to who visited England to see whom about the Graham orrery. In addition to the above story the brilliant French scientific instrument and clockmaker Philip Vayringe, having been appointed to the royal position of watchmaker and mechanist at Luneville, visited the same place as Eugene to collect instruments of the orrery kind. He even resided in London around that time with the eminent Desaguliers while he learned more about the model's mechanism, possibly from Graham himself.

Vayringe was also commissioned to build more such 'planetaria' as well as repair an 'English model' in the royal collection. In particular he was commissioned to build a copy of Graham's clockwork design. He produced one with just a few modifications and with an elaborate case and pedestal that pleased Duke Leopold of Lorraine so much that he had Vayringe present it personally to the Archduchess Maria Theresia. It can currently be seen in the Naturhistorische Museum in Vienna. In 1725 he worked on his own complex orrery with many planets and satellites that remained unfinished, and more. Vayringe died in 1746 of malaria.

Chance has clearly played a hand in the naming of these geared models of the Solar System. They could easily have been named rowleys or eugenes or even savoys, although the latter might cause some confusion with cabbages!

Much of the story of clocks mingled with royalty that constantly changed. Table 2.1 shows monarchs of the era as a reference while following the trail.

Table 2.1 Select list of Monarchs of the Fifteenth to Eighteenth Centuries

Monarch	Royal House	Born/Died	Reigned	Notes
Henry VIII	Tudor	1491–1547	1509–1547	Noted for his battles with the Church and his six wives
Edward VI	Tudor	1537–1553	1547–1553	Nine years old when he succeeded to the throne and only 15 when he died, probably of tuberculosis
Lady Jane Grey	Tudor	1537–1554	1553	'Nine-Days Queen', executed at age 16
Mary I	Tudor	1516–1558	1553–1558	Known as Bloody Mary, because 300 protestant martyrs were burned at the stake. Married Philip II of Spain. Died of influenza
Elizabeth I	Tudor	1533–1603	1558–1603	The Virgin Queen. Defeated the Spanish Armada
James I of England and James VI of Scotland	Stuart	1566–1625	1603–1625	Married Anne of Denmark. King of England, Scotland, France and Ireland
Charles I	Stuart	1600–1649	1625–1649	Charles dissolved Parliament for 11 years Executed in 1649
Interregnum	Common-wealth Protectorate		1649–1653 1653–1659	Empowered by Parliament Led by Oliver Cromwell
Charles II	Stuart	1630–1685	1660–1685	Restoration. Recovery from Puritanism. Married Catherine of Braganza
James II	Stuart	1633–1701	1685–1689	Married Italian princess, Mary of Modina. Ousted in 1689

Table 2.1 (continued)

Monarch	Royal House	Born/Died	Reigned	Notes
William and Mary	Orange Stuart	1650–1702 1662–1694	1689–1702	William III (II of Scotland) was a Dutchman 'invited' to England. Mary was the daughter of deposed James II
Anne	Stuart	1665–1714	1702–1714	Married Prince George of Denmark. No surviving heirs. Queen of Great Britain, France and Ireland
George I	Hanover	1660–1727	1714–1727	German. Great grandson of James I. Married cousin Sophia Dorothea of Celle
George II	Hanover	1683–1760	1720–1760	King of Great Britain, France, Ireland. Duke of Brunswick-Lüneburg, Elector of Hanover. Married Caroline of Ansbach
George III	Hanover	1738–1820	1760–1820	King of the United Kingdom of Great Britain and Ireland. Married Charlotte Sophia of Mecklenburg-Strelitz. Burdened by long bouts of 'madness'
George IV	Hanover	1762–1830	1820–1830	Married German princess, Caroline of Brunswick. His only daughter, Princess Charlotte, died young
William IV	Hanover	1765–1837	1830–1837	Age 64 at succession. Married Adelaide, daughter of the Duke of Saxe-Meiningen. No surviving legitimate heirs
Victoria	Hanover	1819–1901	1837–1901	The longest reign in British history. Ruled the biggest empire the world has ever seen

3. Orrery—the Man and the Model

The Boyle Family

The popularity and educational importance of the tellurium or planetarium owes a great deal to Charles Boyle, the fourth Earl of Orrery. He allowed his name and title to be adopted for the geared model of what was, initially, a small part of the Solar System. But who were the Orreries who accelerated such learning and awareness? The family was the Boyle family, who firstly acquired the title of Earl of Cork and later the title of Earl of Orrery. As the titles passed down through the ages, they were combined.

The history and achievements of the distinguished Boyle family and its influences could and has filled many volumes, but a brief overview is warranted here. In particular we give the names and a few notes on the children of each relevant Boyle. They are numerous, and some names are repeated, which can be confusing, such as the Rogers and Richards. Within the notes are references to family names such as Sackville that identify links to places such as Knole in Sevenoaks, Kent. The diagram (Fig. 3.1) for the sixteenth to eighteenth century Boyles should help as a reference while following the events.

In 1066, men from Normandy, led by William the Conqueror (Fig. 3.2), conquered the British Isles, killing, maiming and subjugating the existing population. It would be reasonable to assume, thus, that a large portion of indigenous Britons derive from the Normans. It is more than likely that the Anglo-Norman knights, the de Beauvilles (de Boyville), initially settled in Wales and northern England. Some traveled to Ireland and some made their way to Scotland. These last included Clan Boyle, probably with the Norman knight Hugh de Morvile, in the early twelfth century. This led to the two great Boyle families, although, from thereon, they have separate genealogy.

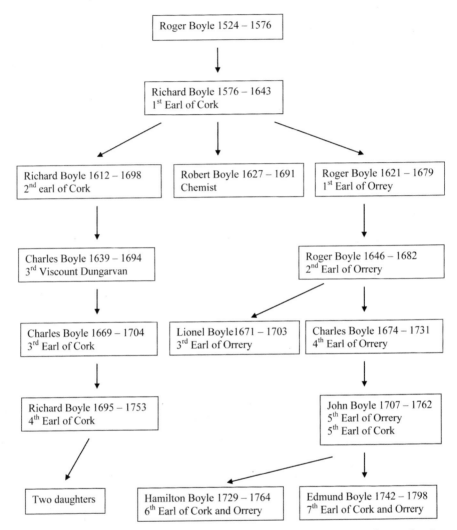

FIG. 3.1 A portion of the Boyle family tree. (Illustration by the author)

The original family name probably came from the old Irish word *Baigell*, meaning 'having profitable pledges' (Irish *geall* means 'pledge'), which turned into Baoighill, then into the O'Boyles, who were a large group that settled in County Donegal. Ballyweel, in County Donegal, is a phonetic contraction of Baile Ui Bhaoighill, which means the home of the Boyles.

To approach the history relevant to clockmaking and telluriums, an appropriate start is to leap forward from the Normans to when the Boyles spread out from Ireland all over the world,

Fig. 3.2 A print from an early engraving, 1745, of William the Conqueror. (Courtesy of the Trustees of the British Museum)

but particularly to England. Richard Boyle, the first Earl of Cork, Roger Boyle, the first Earl of Orrery, Dr. Robert Boyle, the great scientist and Charles Boyle, the fourth Earl of Orrery, are all an important part of the story.

Richard Boyle

Richard Boyle, the first Earl of Cork, was otherwise referred to as the Great Earl of Cork and was very much the founder of all that was successful in the family's future. He was also the first Viscount Dungarvan and Baron Boyle of Youghal. His father, Roger Boyle (1524–1576), was born in Herefordshire but lived in Kent, England, and died in Preston, a village on the outskirts of Faversham. He and his wife are interred in the local parish church of

FIG. 3.3 Roger Boyle died in Preston, a village on the outskirts of Faversham, Kent. He and his wife are interred in the local parish church of Preston. (Photograph by the author)

Preston, within which is an impressive marble tomb erected by Richard Boyle, first Earl of Cork with the children John, Hugh, Elizabeth and, of course, Richard himself.

Although the roots of this beautiful church go back to Anglo-Saxon times, nothing remains of these, and it was in the time of the Normans, from around A.D. 1100, that lasting structures were built. Figure 3.3 shows the picturesque Preston church.

Richard Boyle was born in Canterbury and attended King's School at the same time as Christopher Marlowe. Both also attended Corpus Christi College, Cambridge. Boyle went to London to study law, but left in 1588 before completing his studies to go to Dublin, Ireland. He held many offices of state, such as Deputy Escheater (who deals with the property and belongings of those who die intestate), and was Member of Parliament for Lismore.

However, Boyle had been the object of attacks by Sir Henry Wallop (Treasurer at War), Sir Robert Gardiner (Chief Justice of the King's Bench), Sir Robert Dillon (Chief Justice of the Common Pleas) and Sir Richard Bingham (Chief Commissioner of Connaught). This was a demonstration, said Boyle, of their envy of his success and increasing prosperity.

Boyle was arrested on charges of fraud and collusion with the Spanish (essentially accusations of covert papist infiltration, a treasonable offence for an official in Elizabeth I's Protestant civil service). He was thrown into prison several times during this period, at least once by Sir William FitzWilliam in about 1592. He was about to leave for England to justify himself to Elizabeth when there was a rebellion in Munster in October 1598, in which "all my lands were wasted", and once again he was returned to poverty. The Nine Years' War arrived in Munster with Irish rebels from Ulster, who were joined by locals who had lost land to English settlers. Boyle was forced to flee to Cork for safety.

This turn of events left him obliged to return to London and his chambers at the Temple. At this point he was almost immediately taken into the service of Robert Devereux, second Earl of Essex.

Henry Wallop then renewed his prosecution of Boyle. Boyle was summoned to appear at the Court of Star Chamber. In the proceedings, Boyle's adversaries seem to have failed to substantiate their accusations. Boyle had somehow managed to secure the attendance of Elizabeth herself at the proceedings, and he successfully exposed some misconduct on the part of his adversaries. For some time he was in and out of poverty and prison due to the vagaries of politics and religion before fleeing to Cork and then England where he, again, struggled to keep himself out of prison probably due to the jealousy of his adversaries.

However, Boyle was favored by Queen Elizabeth against his enemies and so was appointed clerk of the council of Munster in 1600. He was knighted in 1603, the same day that he married his second wife, Catherine (1582–1630). His first wife, Joan Apsley, was 17 years young at marriage to his 28, but she died in childbirth (the son was stillborn) in 1599 and was buried in Buttevant Church, County Cork. Her dowry provided him with a good income for many years, £500 per year, equivalent to around £ 70,000 by today's values. His second wife Catherine was of high status, as she was the daughter of Sir Jeffrey Fenton, Principle Secretary of State and Privy Councillor in Ireland.

There were 15 children from the marriage. Some died young, but several lived to a good age for the time. Those who were important for the story of the orrery are:

Sir Richard 'The Rich' Boyle: 1612–1698, second Earl of Cork, first Earl of Burlington, Lord High Treasurer of the Kingdom of Ireland, Viscount Boyle of Kinalmeaky, Baron of Bandon Bridge, first Baron Clifford of Lanesborough in the county of York. Born in Youghal and buried in Lanesborough, Yorkshire, England.

Sir Roger 'The Wise' Boyle: 1621–1679, first Earl of Orrery, Baron of Broghil, Lord President of Munster. Born in Lismore, County Cork, died in Castlemartyr. Buried in Youghal.

Robert 'The Philosopher' Boyle: 1627–1691. Born in Lismore, buried in St Martin-in-the-Fields, Westminster, England.

All of Boyle's positions of state, acquisitions and honors are too numerous to be mentioned here, but a couple are important to give full credit to him. He bought the entire estate that belonged to Sir Walter Raleigh, the explorer who eventually had his head chopped off. These estates amounted to 42,000 acres in the county of Cork and included the Youghal peninsular, as well as Castlemartyr, Charleville, Doneraile and Midleton towns. He became Lord Boyle, Baron of Youghal in 1616 and Earl of Cork and Viscount Dungarvan in 1620. Order on the Boyle estates was maintained by his men in 13 castles. In 1631 he became Lord Treasurer of Ireland. He had many manors and estates in the southwest of England, where he was delighted to have good access across the water to Youghal. Youghal is a most attractive country town and Fig 3.4 shows a colorful view across the Blackwater River.

At Youghal is the ancient Collegiate Church of St, Mary, which dates back possibly to A. D. 450 and definitely to 1170, as confirmed by carbon dating of the roof timbers (Fig. 3.5). The Earl of Cork had much to do with restoration of this church from the time of his ownership in 1606, and he built a huge monument within to commemorate his family as it was at the time. There is a real feeling for the Boyle history while standing before this mag-

FIG. 3.4 A colorful view at night over the river Blackwater toward Youghal, Ireland. (Photograph by the author)

FIG. 3.5 The Collegiate Church of St. Mary at Youghal. (Photograph by the author)

nificent commemoration (Fig. 3.6) and the detail shown (Fig. 3.7). The residence of Sir Walter Raleigh, Myrtle Grove, still exists adjacent to the Collegiate church, although it is currently in a dilapidated state and awaits restoration (Fig. 3.8).

FIG. 3.6 A magnificent commemoration within The Collegiate Church of St Mary. (Photograph by the author)

FIG. 3.7 A detail of the commemoration within The Collegiate Church. (Photograph by the author)

FIG. 3.8 The residence of Sir Walter Raleigh, Myrtle Grove adjacent to the Collegiate Church. (Photograph by the author)

Roger Boyle

Roger Boyle was created the First Earl of Orrery on September 5, 1660, just 3 months after the restoration of the monarchy, with Charles II returning to Britain. The name Orrery possibly came from the name of an Irish tribe called the Orb's People or, in Gaelic, Orbhraighe. The land they occupied became a territory, then a barony. The town of Charleville in north County Cork was founded in 1661 by Roger Boyle, who named it after the new King Charles II to prove his loyalty to the crown. The villages of Brohill and Rathgoggin preceded the formation of the town of Charleville in the area. It is in the parish of Rathgogan, barony of Orrery and Kilmore, county of Cork and province of Munster.

He was engaged in Royalist schemes, however, when Oliver Cromwell visited him and explained that he knew all about his activities. Cromwell offered him the choice of clearing himself by serving the commonwealth in Ireland or be sent to the tower. He accepted and served Cromwell throughout the campaign. He was largely responsible for the return of the Monarchy and consequently was taken into great favor when Charles II became king. In addition to Lord Orrery's achievements as a statesman and administrator, he gained some reputation as a writer and dramatist. He died on 26 October 1679.

Roger Boyle was married to Lady Margaret Howard, who was born in 1622 and baptized at St Martin-in-the-Fields Church, London. She was buried at Isleworth, Middlesex, England in 1689.

There were, possibly, seven children, although the lives or fates of three daughters are unknown. The two sons are:

Roger Boyle: 1646–1682, 2nd Earl of Orrery. Married Lady Mary Sackville in London in 1664 (relevant to Knole House and 4th Earl of Orrery).

Henry Boyle: Born 1648 in Castlemartyr, Cork, and died in 1693 fighting in the Duke of Marlborough's campaign in Flanders, France, possibly with Prince Eugene of Savoy, who purchased instruments from George Graham.

Robert Boyle

Robert Boyle, fourteenth child and sixth son of The Great Earl of Cork, became a renowned scientist. Much information is given on a plaque in the Collegiate Church of St Mary. It reads:

Robert Boyle,
Natural philosopher and founder of modern Chemistry.
Seventh son of Richard Boyle first Earl of Cork.
Born in Lismore castle on January 25th 1627.
As a child learned to speak Latin and French.
And later Hebrew, Greek and Syriac.
At the age of eight was sent to Eton.
In 1641 went to Italy, and studied under Galileo.
The charter of the Royal Society of London granted by Charles II in 1663 named Boyle a member of the council, and he was elected its president in 1680 but declined.
A complete edition of Boyle's work published in 1772 ran to six volumes quarto.
Boyles law made his name the most celebrated in the English language.
He died on the 30th December 1691.
Boyle was certainly one of Ireland's most celebrated sons.

The imposing Lismore Castle, Robert Boyle's birthplace (Fig. 3.9), was previously owned by Sir Walter Raleigh and sold to Richard Boyle whilst Raleigh was imprisoned for high treason.

FIG. 3.9 Lismore Castle, Ireland, birthplace of Robert Boyle the chemist. (Photograph by the author)

To progress his interest and work in chemistry, he moved in 1654 from Ireland to Oxford, England. With the assistance of Robert Hooke, he worked on an improvement to Otto von Guericke's air pump to produce a Pneumatical Engine. This was completed in 1659. In 1661 he published his first major work, 'The Sceptical Chymist', which is acknowledged as a turning point in the development of chemistry.

Boyle's Law is surely one of the most recognizable formulae in science, especially as taught in schools. It shows the relationship between the pressure and the volume of a gas, PV = a constant, and Boyle is rightly credited with its popularization. However, this is not the whole story. Richard Towneley was an English astronomer and mathematician and one of a group of pioneering astronomers that included Jeremiah Horrocks, William Crabtree and William Gascoigne. Towneley, with his physician and friend Henry Power, discovered a relationship between the pressure and the volume of a gas in an enclosed system. They then corresponded with Robert Boyle, who carried out more investigations and brought this to more popular attention with his writings. He initially referred to it as Mr Towneley's hypothesis. By all accounts there was no animosity or credit-grabbing, they just all worked together in science. The literature is somewhat contradictory as to whether Towneley or Power first put forward the hypothesis. It was while Boyle was answering objections to some of his experiments by Francis Line

(1595–1675) that Boyle first mentioned the law. The likely truth is that Power and Towneley carried out some gas pressure/volume experiments, but it was only Boyle, with the constant assistance of Robert Hooke, who thoroughly tested the relationship and proved the law beyond doubt. But for a printer's error (duly acknowledged by all), the publication of findings leading to the law would have given equal credit to Hooke and maybe even co-authorship. From Hooke's diaries it is clear that he was often in the company of Robert Boyle and also with Tompion with regard to a barometer.

In 1668 Roger Boyle travelled to London, where he remained for the rest of his life. He lived with his older sister, Lady Ranelagh, (Catherine Boyle, 1614–1691), who also helped him with his experiments, for which she provided a laboratory within the house. They both died in the same year, Robert just a week after Catherine.

His achievements in science were vast! As the acknowledged founder of chemistry, he really understood particles, substances and analysis. He set 24 goals for the world to achieve, many of which have already been solved. These include the prolongation of life, pain relief, drugs to alter perception, a certain way of finding longitudes and a ship that could sail with all winds. He also carried out much work on the defining and spreading of Christianity.

Charles Boyle, fourth Earl of Orrery

The fourth Earl of Orrery was born in Little Chelsea, London, in July 1674, the youngest son of Roger Boyle, second Earl of Orrery, and Lady Mary Sackville. His grandfather was Roger Boyle, the first Earl of Orrery, his great-uncle was Robert Boyle, the natural philosopher described above.

His parents separated and he was brought up by his mother at Knole, Sevenoaks, Kent, which was the home of his uncle, Charles Sackville, sixth Earl of Dorset. The stunning façade and court colonnade of his house are shown in Figs. 3.10 and 3.11. Charles Boyle was educated at a Sevenoaks school, which was housed in small buildings around the town (even outside the town in the 1730s) until a permanent schoolhouse was built in 1730 to the designs of Lord Burlington, a friend of the then headmaster Elijah Fenton. A diagram of the links between the Boyle and

Fig. 3.10 Knole House, Sevenoaks, Kent. (Photograph by the author)

Fig. 3.11 The stunning colonnade at Knole House. (Photograph by the author)

110 Orrery

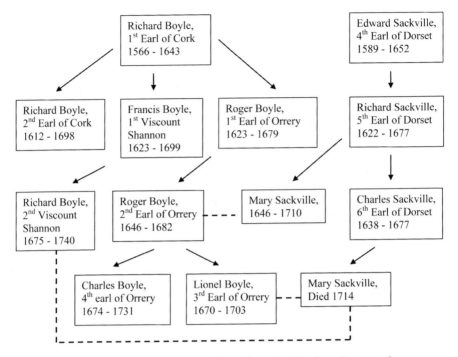

Fig. 3.12 Links between the Boyle and Sackville families in the seventeenth to eighteenth centuries. (*Solid line*: Lineage, *Dotted line*: Marriage) Mary (1) Sackville, married to Roger Boyle, second Earl of Orrery, would be the mother of Charles, the fourth Earl of Orrery. Mary (2) Sackville, wife of Lionel, third Earl of Orrery, was the illegitimate daughter of sixth Earl of Dorset. Mary (2) Sackville's mother-in-law, Mary (1) Sackville, was Charles, sixth Earl of Dorset's sister. The second marriage of Mary (2) Sackville was to her first husband's uncle, Richard Boyle, second Viscount Shannon.

Sackville families is shown in Fig. 3.12. Charles then went to St Paul's in London before attending Christ Church College, Oxford, where he studied with Francis Atterbury, later Bishop of Rochester. Atterbury could have been a most unfortunate influence on the young Charles at the age of around 16 since Atterbury was a co-conspirator for the losing Jacobite cause. However, Charles was shown to be most studious and was the only nobleman to take a degree from the college during a 30-year period.

Roger Boyle, second Earl of Orrery, and Mary Sackville separated around 1675. Since their son Charles Boyle was born in 1674, he must have been around 1-year old when he was taken in to be looked after at Knole House, where he spent his youth and attended the school in Sevenoaks, Kent, presumably first under the care and protection of Frances Cranfield with Richard Sackville, fifth Earl of Dorset, then with Henry Powle, the future Master of the Rolls.

Popular history remembers Charles Boyle for honoring the geared model of the Solar System with his name, but delve just a little deeper into his early life at Oxford and you will find him involved in a massive literary battle with one of the most eminent scholars of the time. Under his tutor, Francis Atterbury, he studied not only classical authors but the more modern ones such as Descartes and Locke, finding "a great deal of very good sense" in Locke's "Essay Concerning Human Understanding." The dean, Henry Aldrich, particularly recognized Charles' talents and quoted him to be "the great ornament of our college"! He was a hard worker and, by 1693, had already published translations of Plutarch and Lysander. Because he was considered to be the college's star pupil, he was selected by Aldrich to prepare a work which, by tradition, would be distributed to members as a Christmas gift and would be submitted for publication.

This was the beginning of a heated literary discussion with Richard Bentley, the Master of Trinity College, Cambridge, which did not influence Boyle's future, but is certainly remembered as part of the continuing contests between Cambridge and Oxford.

The text chosen was The "Epistles of Phalaris," of which there were 148. The letters were a particularly common topic of the time to discuss, the controversy being whether they were written by Phalaris or by someone else. An idle comparison between ancient and modern learning, begun in France, had spread to England. Sir William Temple, then eminent as a man of letters, published an essay in 1690 in which he gave the preference to ancient literature, in general, and praised the letters of Phalaris, in particular, as superior to anything since written of the same kind.

The semi-legendary figure of Phalaris (c. 570–554 B.C.) was born in Astypalaea, a city of Crete and later ruler of Agrigentum in Sicily, where he became a tyrant. The image (Fig. 3.13) represents Phalaris condemning the sculptor Perillus to the torture device named the Brazen Bull, after Baldassare Peruzzi. Some historians add a kinder aspect to Phalaris, citing him to be a supporter of learning.

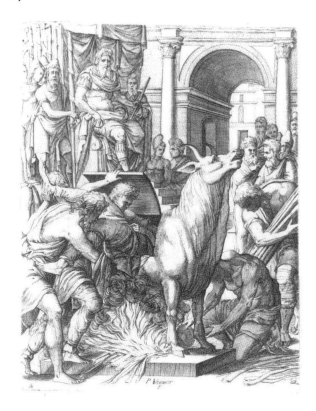

Fig. 3.13 An engraving, copperplate print of Phalaris condemning the sculptor Perillus to the Brazen Bull, after Baldassare Peruzzi, before 1562

So, did Phalaris write the letters or did someone else? Charles thought that they were genuine and that formed the basis of his text in 1695. In Charles' opinion some research information had been withheld from him, a manuscript in the king's Library, "pro singulari sua humanitate," by Dr. Richard Bentley, a Master of Trinity College, Cambridge, and a jibe against the doctor was included in his dissertation. Dr. Bentley was stung by the insult from an undergraduate and, turning his mind to the Epistles, produced a masterful dissertation showing them to be spurious. It is interesting to note that it was, in fact, Boyle's editor, in his preface, that added an insulting reference to Bentley and complained of his discourteous conduct in refusing the use of a manuscript of Phalaris kept in the royal library. Bentley wrote at once to Boyle explaining that there had been a mistake and that he had intended no discourtesy; but Boyle, acting on the advice of others, refused to make any amends. His reply was practically a provocation for Bentley to do his worst. For such an eminent scholar it is interesting that his name should crop up so frequently in connection with a spat with Charles Boyle.

Charles neither wished to stay at Christ Church College, nor did he plan to return to Ireland. He succeeded only in securing a minor sinecure in Chancery rather than his preference, at that time, as groom of the royal bedchamber. He was Receiver-general in the alienations office from 1699 to 1717 and Gentleman of the Bedchamber 1714 and 1716. Deprived of the opportunity to impress at court, he pursued his literary ambitions, which included writing the comedy *As You Find It*. He joined the coffee-house circle of John Dryden (1631–1700), recognized as one of the greatest literary figures of his age. Charles' literary efforts did not always go well as is particularly exemplified by Lady Mary Wortley Montague's devastating dismissal of his literary pretensions, but she was probably also motivated by family animosities.

Although Charles spent much of his time caught up in the literary controversy with Richard Bentley of Cambridge, he also developed a wide-ranging intellectual curiosity that resulted in a huge library of 10,000 volumes of medical and scientific works ranked as the finest in England. These included works in Greek, French, Italian and Latin as well as a complete set of 'The Journals of the House of Lords'. He bequeathed the library, along with his scientific instruments, to Christ Church Library (now on display in the Oxford Museum of the History of Science). So important was the collection that the Oxford museum writes: "The instrument collection of Charles Boyle, fourth Earl of Orrery (1676–1731), could keep this column supplied with spheres of different kinds for many years to come. Bequeathed to Christ Church and placed on loan to the Museum by the College, it is one of the most important surviving examples of a cabinet of instruments, coming from a period when one manifestation of a fashionable interest in natural philosophy was the assembling of just such collections by wealthy amateurs."

Boyle became a Fellow of The Royal Society in 1706. One might have assumed that since his great-uncle was the eminent scientist Dr. Robert Boyle, the library collections would have included some of his works and instruments. But that is not the case. In fact, the paths of nephew and uncle rarely crossed. It appears that Charles' interest in science was the reading and collecting of books and the amassing of scientific instruments, not using them or creating any of his own. Although he was elected FRS, he reportedly never attended.

Boyle's political career began by being elected to the family borough seat at Charleville, County Cork, which he represented jointly with John Ormsby from 1695 to 1699. Once again, controversy followed him, having been elected to one of the borough seats for Huntingdon. Apart from having charges of impropriety against him withdrawn, the confrontation with his fellow MP Francis Wortley did not end well, since Wortley challenged him to a duel in Hyde Park. After a fierce struggle Charles was badly wounded (although Wortly conceded) and took several months to recover. In 1703 he became the fourth Earl of Orrery on his brother's death. After generally 'toeing the line' politically, he continued to be elected to the Huntingdon constituency before purchasing a colonel's commission with three major foot regiments and was awarded the honor of a Knight of the Thistle. He was an opponent of the Tack, a sublease of a portion of the land that had been granted by the monarch to a major landholder.

Charles married Lady Elizabeth Cecil in 1706, who died in 1708 just a year after the birth of their son John Boyle, later fifth Earl. John was born at Charles' house in Glasshouse Street, Westminster, and renting records indicate that Charles retained these lodgings from 1717 until his imprisonment in 1722. FRS records cite an address in the adjacent Vigo Lane in 1718. He inherited the estate at Marston Bigot in Somerset in 1714, which he used as a summer retreat. Later he purchased a house nearer London, Britwell House in Burnham, Buckinghamshire. His last home was in Downing Street, Westminster. The rate books from 1724 to 1732 inclusive show the Earl of Orrery at the house, after which it was empty for 6 years. The fourth Earl died on 28 August 1731, so that it is possible that his son lived there for a short time. Also quoted is that the first occupant of the house forming the western portion of the present No. 11, Downing Street, was the Earl of Orrery.

He never married again after his wife's death but had two sons and two daughters with his mistress and wife of his secretary,

Margaret Swordfeder. He named the sons Charles and Boyle and the two daughters Clementina and Martha Sophia, which hinted at Charles' underlying Jacobite sympathies. Charles fell out with his son John over the affair and therefore excluded John from his will. Although father and son made up just before the death of Charles the will stood and, apart from scientific collections, was in favor of Swordfeger.

As with many of the Boyles, Charles was a fighting man and during 1709 and 1710 served in three military campaigns in Flanders. He was involved in the fierce fighting at Malplaquet in 1709 and at the siege of Béthune. As a major-general fighting with the prestigious Royal Scots Fusiliers, he advised the first minister, Robert Harley, on military reforms designed to offset the influence of their mutual political opponent the Duke of Marlborough. From 1711 to 1713 he served as the queen's envoy-extraordinary at Brussels and The Hague.

His growing influence with ministers and at court brought him a place on the Privy Council (1711) and a seat in the House of Lords as Baron Boyle of Marston, Somerset. However, he persisted in his leanings toward the Jacobite cause, which eventually led to his imprisonment in the Tower for several months in 1722–1723. He was released only for a sum of £ 50,000 at the pleading of his family, who were concerned for his health. He continued to blatantly support the opposition in parliament as well as indulge in Jacobite activities. During preparation for one of such activities he died at his home in Downing Street, London, 1731, and was buried at Westminster Abbey.

The stunning portrait of Charles Boyle, fourth Earl of Orrery, is shown in Fig. 3.14.

Budgell's Boyle

Historians all have their take on the details of the life of Charles Boyle, but who better to consult than someone who was actually there! Eustace Budgell was his friend and confidant and discussed all manner of topics while "walking across the park with him from his own house". But care must be taken to treat as fact everything a close and admiring friend has to say, especially one with such a mixed history of vanity and vindictiveness.

Fig. 3.14 Portrait of Charles Boyle, fourth Earl of Orrery. (Courtesy of National Portrait Gallery, London)

Eustace Budgell (Fig. 3.15) was born in 1686 in Exeter and died in London in 1737. He was a writer and essayist and a major contributor to 'The Spectator' periodical, published by Sir Richard Steele and Joseph Addison, while also contributing to the 'Tatler' and 'The Guardian'. His cousin Addison, then secretary to the Lord Lieutenant of Ireland, gave Budgell a clerkship in 1710, which he carried out with great ability in addition to other posts. From 1718 things began to go badly. Addison resigned due to ill health (he died the following year, 1719), and Budgell quarreled with his successor the Duke of Bolton, which led to Budgell's dismissal. He lost £ 20,000 in the notorious investment disaster of the South Sea Bubble and became disliked because in his writings he attacked eminent people such as Sir Robert Walpole. He was

FIG. 3.15 Eustace Budgell, friend and confidant of Charles Boyle. (Credit: PD-US, obtained from Wikipedia)

criticized in particular by Alexander Pope in the 'Epistle to Dr. Arbuthnot' and in 'The Dunciad'. When his life became intolerable, probably because of litigation concerning the forging of a will to make himself a beneficiary, he filled his pockets with stones and drowned himself.

The year 1703 was a great year for Charles: the title Earl of Orrery was bestowed upon him. In 1706 he married Lady Cecil, daughter of John, Earl of Leicester. He worshipped her and always referred to her with great affection. He was understandably devastated when she died just 2 years later, not long after the birth of their son John.

In war Charles acted with the utmost gallantry. According to a personal assurance by an officer serving under Charles, during a particularly bloody battle Charles fought at the front where the action was greatest and where most of his men fell all around him. He was not parasitic or excessive with expenditure of the public purse. When he was Envoy Extraordinary for the Queen, he carried out his duties most diligently although he assured Budgell personally that he never received a single farthing from the

Treasury all the while he was in Brussels. Budgell asserts "He maintained the dignity of his post in every respect, kept a most elegant table and made himself universally loved and esteemed by those who had any transactions with him". He was loyal to his friends, to whom he was always at home and never denied, but wouldn't tolerate those he despised. He arose early, by 6 am, and was on horseback by 7 am unlike, according to Budgell, most of the other English nobility.

There have been many suggestions that Charles was involved in secretive Jacobite plots; these are part of the history learned. However, Budgell's accounts do seem to point to the opposite. Charles had jealous enemies at court who used enactment of a law that allowed suspension of Habeas Corpus (no imprisonment without a fair trial) that held for a year. Charles was thrown into prison, the Tower, without any recourse to justice or a fair trial. A prisoner could be treated in any manner deemed to be fit by an unscrupulous minister. Charles was treated badly, so badly that his health deteriorated seriously. It is pertinent to reflect on Boyle's probable involvement in the Atterbury Plot.

The plot centered on the Bishop of Rochester, Francis Atterbury (1663-1732), with whom Charles studied in his early years at Oxford, a serious attempt to restore the Stuart monarchy in Britain. In addition to Atterbury, Charles Boyle, Lord North and Sir Henry Goring were involved, but the most active Jacobites were John Plunket, George Kelly and Christopher Layer. The documents used by the investigating parliamentary committee against them came from the personal archive of Sir Robert Walpole, British Prime Minister at the time. Having unraveled the plot, Atterbury lost his bishopric and was exiled, John Plunket spent the rest of his life in the Tower of London and Christopher Layer was hung, drawn and quartered at Tyburn on 17 May 1723. He was possibly more severely punished because he refused to inform on his accomplices.

Budgell writes that no evidence implicating Charles was uncovered however hard the state representatives delved, and Charles was therefore found innocent of all charges. Because of Boyle's dwindling health, which had brought him almost close to death, he was released on the payment of £50,000, as described above. It would be easy to conclude that Budgell was just being loyal and wished to exonerate his friend from any ill-doing. His attitude might also explain the absence of any specific reference to Charles' relationship with Margaret Swordfeger, except to mention that he, as all aristocracy, involved themselves with other women. Further confirmation of Budgell's close friendship is that he was one of the few people who had a key to Charles' library and could let himself in at any time. He describes the library as having three large rooms for French and Italian, English and Greek and Latin authors. Budgell couldn't even concede that the greatest library was Lord Sutherland's, who in his opinion collected rashly and accumulated replicates. Budgell affirmed that Charles almost exclusively never collected any duplicates and therefore had just as large a library as that of Lord Sutherland in the number of 'different' books. He wrote much about how popular Charles was, how religious, or how at least he obliged himself to conform in public to the established religion. Budgell also heaped praise on Charles for making up with his son, John, so much so that any observer would think that they were very best buddies and went everywhere together.

In spite of much obvious bias in the writings of Budgell, the tome contains much that should be taken into consideration in the life of Charles Boyle, fourth Earl of Orrery.

A gallery of orreries

Eighteenth-century models

Graham and Tompion really started something and, accelerated by Rowley and the Earl of Orrery, the production of orreries increased enormously throughout the eighteenth and nineteenth centuries. The best clockmakers were eager to build the fashionable models, and the rich, royal and famous were eager to buy them.

The whereabouts of some of the first orreries has been covered in a previous chapter. The first experimental model made by George Graham is in the Chicago Adler planetarium and is shown in Fig. 3.16.

It is amazingly fortunate that the very first experimental model has been recovered and is in safe hands. Prototype or ex-

FIG. 3.16 George Graham proto-orrery, 'G' model. (Courtesy of the Adler Planetarium & Astronomy Museum, Chicago, Illinois)

perimental, typical of the day is that it is still beautifully neat and tidy and in a handsome case.

The next model, made by Tompion and Graham, is housed in the Oxford Museum of the History of Science, and what a great tale has been told regarding the fortuitous destiny of the precious model (Fig. 3.17)!

Zooming in on the detail (Fig. 3.18) reveals how simple the visible part of the astronomical workings appears, and yet it was such a leap forward for the design of a new scientific instrument! The big difference, of course, between the G and the G&T model was that the first was powered by clockwork with eight-day going and fusée and the second required manual turning of a handle. The handle is more easily noticed in the expansion (Fig. 3.19).

There are two pointers indicating dates on the calendar scale around the edge. They are set 11 days apart and refer to the Julian and Gregorian calendars, respectively. While the Gregorian cal-

Fig. 3.17 The second proto-orrery signed by George Graham and Thomas Tompion, (G&T model). (Courtesy of the Museum of the History of Science, Oxford)

Fig. 3.18 The visible part of the astronomical workings of the G&T model. (Courtesy of the Museum of the History of Science, Oxford)

endar was used in the continental countries of Europe, it was not adopted in England till 1752.

The fascinating example shown in Fig. 3.20 is the famous one made by the London instrument maker John Rowley, which started the 'orrery' label for mechanical planetary models. Just as for the G&T model, it is a demonstration device to show the motions of the Earth and Moon around the Sun. These devices became ever more popular during the eighteenth century, especially following on from the universal theory of gravity published

122   Orrery

FIG. 3.19 The handle of the G&T model indicating the manual rather than clockwork mechanism. (Courtesy of the Museum of the History of Science, Oxford)

FIG. 3.20 John Rowley model that started the 'orrery' label for mechanical planetary models. (Courtesy of the Science Museum/Science & Society Picture Library)

in 1687 by Sir Isaac Newton (1642–1727). It is gratifying that this particular example, dated 1712–1713, made for Charles Boyle, the fourth Earl of Orrery, was so colorful and appealing that it de-

FIG. 3.21 The beautifully artistic Rowley improvements are highlighted in the detail image. (Courtesy of the Science Museum/Science & Society Picture Library)

served the noble name that fell off the tongue so easily. The G&T model was soon to be on its way to Prince Eugene, therefore Rowley copied it to present it to his sponsor Charles Boyle. Actually, it might have been presented to his son John, fifth Earl of Cork and Orrery. Although looking very different, the new model still only depicted the Sun, Earth and Moon. Museum records indicate that the object was formally acquired in 1952 from the fourteenth Earl of Cork and Orrery. It had spent some time within the museum's collection as a loan object, but was purchased by the museum for £29,000. Museum records state that an article dated 18 December 1937 in the Illustrated London News confirms the orrery to be in the possession of the Earl's family since 1716. The beautifully artistic improvements are shown in the detail image Fig. 3.21. Standing in front of the Rowley orrery in the Science Museum, one can see other original accessories of the model, removable, beautifully made brass sleeves with glass domes possibly for the protection of the model's components during transportation (Fig. 3.22).

Throughout the story of the birth of the orrery, those few hundred years ago, there has been uncertainty about some facts, such as who named the model 'orrery'. Trivial facts they might be, but while telling the story, one is transported to those past times to recreate, and therefore follow and appreciate, the lives and practices and the trials and tribulations of the dedicated and driven brilliant artificers. Rupert Thomas Gould is a more modern genius

Fig. 3.22 Removable, beautifully made brass sleeves with glass domes can be seen on the model on display at the museum. (Photograph by Tony Buick with permission of the Science Museum/Science & Society Picture Library)

who probably knew more about the Rowley orrery than anyone else. He researched the model in intricate detail, sufficient to repair and refurbish it from its somewhat 'stored state'. He should, therefore, have the last word. But Gould has much more to do with the history of clocks and orreries than just this present question.

Rupert Thomas Gould (1890–1948) was in the British Royal Navy as a navigator and rose through the ranks to achieve the title of Lieutenant Commander. He became an expert in naval history, cartography and expeditions to the polar regions. His brilliance in his amateur obsession with clocks made him an acknowledged expert in horology, and his book 'The Marine Chronometer', published in 1923, remains the definitive work even after all these years. He is perhaps most popularly known for his restoration of the original Harrison clocks that were so much part of the story of longitude, made famous in Dava Sobel's story 'Longitude' and the film of the same name.

Gould was entrusted to overhaul the Rowley orrery by the Admiral, Earl of Cork and Orrery, Commander-in-Chief at Portsmouth. In the article 'The Original Orrery Restored' in The Illustrated London News, December 18, 1937, he describes the restoration and the mechanism: "On turning the handle—one turn corresponds to the passing of one day—the following motions occur: the Sun, whose axis is correctly inclined to the Ecliptic (about 7° inclination, longitude of ascending node, 72°) makes one revolution in its true (equatorial) period—about 27½ days. The earth exhibits its daily rotation, and keeps its axis uniformly inclined at 23½° to the ecliptic" and so on in some detail. Gould acknowledges that the motions of the bodies were reproduced to a considerable degree of accuracy and the construction of the wheelwork must have been very far from simple. While inspecting the orrery, within Gould found a piece of faded paper on which was written "repaired, cleaned and an error in the wheelwork corrected by W.Lovell, April 1876". In fact, Lovell had not corrected anything, but introduced an error! As Lovell had thrown away what he thought to be an erroneous wheel, Gould had to make another to return the mechanism to that of the original and correct version.

As to who named the orrery thus, Gould writes: "At the suggestion of Sir Richard Steele—then in his zenith, essayist, journalist, wit, and man-about-town—the name 'Orrery' was given out of compliment to its noble patron". So there we have it, maybe!

Jesse Ramsden (1735–1800) was one of the most famous English scientific instrument makers for nautical and astronomical use and traded at The Strand in 1763–1766. He invented a semi-automatic dividing machine to mark the divisions on maritime and scientific instruments as well as developing improvements on them. In the second half of the eighteenth century he was approached by Guiseppe Piazzi to create a complete set of astronomical instruments. Piazzi used these to persuade King Ferdinand of Sicily to supply the money to build an observatory at Palermo, Sicily. Ramsden made a great impression on the advancement of astronomy here by creating the Great Circle of the Palermo observatory, the first of a new generation of astronomical instruments with circular scales. Using this instrument to map the stars Piazzi unexpectedly discovered the first asteroid, Ceres.

Ramsden received many awards and much recognition, including being elected Fellow of the Royal Society in 1876. The complex orrery shown in Fig. 3.23 is housed in the London Sci-

FIG. 3.23 Jesse Ramsden's only orrery. (Photograph by the author with permission of the Science Museum/Science & Society Picture Library)

ence Museum. It is believed that this orrery is the only one made by Ramsden and is somewhat incomplete although, because the cover is missing, the inner mechanics can be appreciated.

A new approach to the construction of orreries or planetary models was made by Benjamin Martin in 1764 with the planets held on brass arms fixed to concentric cylinders. However, the system made it impossible to represent all of the Solar System mechanisms and Martin suggested that three separate models should be constructed and placed on a common table and spindle; one for the planets, one to demonstrate the motions of the Moon around the Earth and one to show how the Earth, with its tilted axis, revolved around the Sun. Martin's orrery is shown in Fig. 3.24.

The orrery shown in Fig. 3.25 is housed in the National Maritime Museum, Greenwich Royal Observatory. It was constructed in wood, brass, paper and ivory, and was made by William Jones around 1800. The inclusion of Uranus dates the model to after 1781, when Sir William Frederick Herschel (1738–1822) announced its discovery. The orrery is said to have been owned and used by Margaret Maskelyne, daughter of Nevil Maskelyne, the fifth Astronomer Royal.

Brothers William (1763–1831) and Samuel (1769–1859) Jones were amongst the greatest scientific instrument makers in London, 30 Holborn, in the early nineteenth century. They were also

Fig. 3.24 Benjamin Martin's 1764 orrery with coaxial arms bearing the planets. (Courtesy of Creative Commons Attribution-Share Alike 3.0 Unported, 2.5 Generic, 2.0 Generic and 1.0 Generic license. Attributed to Sage Ross)

Fig. 3.25 A portable orrery on a stand by William Jones. (Courtesy of the National Maritime Museum, Greenwich, UK)

known for marketing globes by W. & T.M. Bardin. The firm was a successor to John Jones & Son. According to some sources, Samuel became active in the business about 1810. The orrery bears the name only of W. Jones. Perhaps William Jones originally made, invented and sold the orrery, or the firm W. & S. Jones sold the model just with the W. Jones name, as originally imprinted.

Also made in 1784 by William Jones, the model shown in Fig. 3.26 was made with wood and surfaced with engraved paper

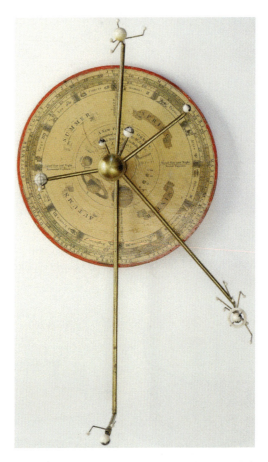

Fig. 3.26 Made by William Jones in 1784, this model included Geordium Sidus, later known as Uranus. (Courtesy of the National Maritime Museum, Greenwich, UK)

bearing calendar and zodiac scales. It also indicates the equinoxes, solstices and seasons. The model includes Geordium Sidus, discovered by William Herschel and later known as Uranus, and therefore confirms a date of post 1781.

All the planets except Earth are made of ivory. It is housed in the National Maritime Museum, Greenwich. Another similar English planetarium, or orrery, also housed in the National Maritime Museum, Greenwich, London, Gabb Collection and made about 1794, has a brass central Sun with the then known planets and known satellites. The printed base paper has zodiac and calendar scales, the seasons and a picture showing the relative sizes of the planets.

Orrery—the Man and the Model 129

FIG. 3.27 The Grand Orrery made in 1780 in England by James Simmonds and T. Malby and co. (Courtesy of the National Maritime Museum, Greenwich, UK)

To illustrate the interaction and confusion commonly encountered of notable historic figures, "William Jones" was (and still is) a particularly common name, especially in Wales. Another William Jones lived at almost the exact time as Graham (1673–1751), 1675–1749, and they must have known each other. He was a Welsh mathematician and most noted for his proposal of the Greek symbol pi to represent the ratio of the circumference of a circle to its diameter. He became a Fellow of the Royal Society in 1711 and was later its Vice-President. He was a close friend of Sir Isaac Newton and Edmund Halley.

As orrery-makers became ever more ambitious, a few models were made that were large and heavy and included many of the solar system objects, not just the Sun, Moon and Earth. They were called Grand Orreries, were ornately designed and hugely expensive. The Grand Orrery shown in Fig. 3.27 was made in 1780 in England by James Simmonds and T. Malby and co. It is beautifully painted and made of wood, metal and brass. Grand orreries could

FIG. 3.28 A detail of Simmonds and Malby's Grand Orrery. (Courtesy of the National Maritime Museum, Greenwich, UK)

only be afforded by the wealthy and were status symbols as much as educational tools. Around the side are painted the signs of the zodiac with a chiming clock between Leo and Gemini. Four movable wooden circles are on the surface as part of getting lots of information out of the model. The legend accompanying the exhibit reads:

"Grand orreries were educational tools and status symbols, affordable only by the most wealthy. This one stands on a polygonal black wooden base with four trestle legs and brass feet, surmounted by a shallow glass dome. Panels around the side of the base are gilt framed and painted with the twelve signs of the zodiac and, between Leo and Gemini, there is a clock signed by the maker (James Simmonds, London), which strikes on the hour. The upper surface of the orrery consists of four moveable wooden circles of different dimensions that are colored blue. This surface conceals all the supports and gearing, made by Malby & Co., in the cavity beneath. Although mechanically driven, the orrery can be speeded up by inserting a handle in a socket and rotating it. Around the outside of the orrery there is a brass ecliptic scale which shows the degrees of the zodiac, the corresponding dates for the Sun's position in the zodiac and the celestial longitude. It represents the relative motions of the six planets (to Saturn) and their known moons around the Sun. Earth (a replacement model) has its single moon, Jupiter has four and Saturn has a ring and seven moons, although these satellites are not mechanically driven. It was possible for the user to set the instrument to a particular date and time so that it would display the planets in the correct position for this moment." Fig. 3.28 highlights some of the intricate detail.

FIG. 3.29 A grand orrery enlarged by Thomas Wright on the instructions of George II. (Photograph by the author with permission of the Science Museum/Science & Society Picture Library)

A Grand Orrery is on display in the London Science Museum and the accompanying note reads:

"In 1733 George II, George III's grandfather, had this orrery enlarged by Thomas Wright, the King's mathematical instrument maker. The original version did not show Saturn, the outermost planet then known. It was one of the most splendid orreries of its time and had cost £1300, the equivalent of 5 years' wages for a skilled worker."

The surround is certainly most ornate (Fig. 3.29) and the astronomical surface plate most intricate (Fig. 3.30).

In Florence in the Istituto e Museo di Storia della Scienza the Venus Orrery designed by James Ferguson is displayed. It was built at the workshop of the Museo di Fisica e Storia Naturale of Florence and is one of the earlier orreries made around 1775 from wood, glass and brass. Operated by a handle, this orrery reproduces the heliocentric movements of Mercury, Venus and the Earth, as well as the geocentric movement of the Moon. James Ferguson (1710–1776) was a Scottish astronomer and physicist who displayed a conspicuous gift for mechanics at a very early age. After becoming a clergyman, he dedicated himself to the observation of the heavens, even building a celestial globe. After many years in Edinburgh he moved to London in 1743. There, he published astronomical tables and lessons. He was elected a Fellow of

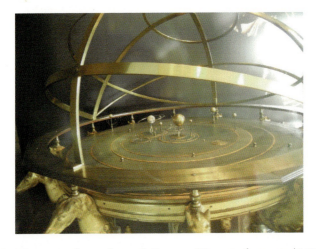

FIG. 3.30 Intricate surface plate of George II's grand orrery (3.29). (Photograph by the author with permission of the Science Museum/Science & Society Picture Library)

the Royal Society of London in 1763. Ferguson described Charles Boyle as the 'inventor' of the orrery!

There were large, grand orreries and portable orreries, but also small ones for the slightly less well heeled. In 1740 Thomas Wright constructed one for 'Ladies and Gentlemen rather than Noblemen or Princes'; this is displayed in the London Science museum (Fig. 3.31).

The accompanying note reads: "It belonged to a science lecturer, Stephen Demainbray, whose collection became amalgamated with the royal collection. It shows the Sun, Earth, Moon, Mercury and Venus. Earth can be turned on its axis directly while the handle is used for showing annual rotation. The outer ring shows both the old and new calendars, the signs of the zodiac and the points of the compass." A detail is shown in Fig. 3.32.

Another beautiful orrery by Ferguson built around 1760 is housed in the same museum. It is made almost entirely of wood and is driven by strips of rope passing through winches. It shows the Earth's yearly revolution around the Sun, the oscillation of the Moon's orbit, the phases of the Moon and the movement of the lunar apogee. The circles around Earth are calibrated to show the signs of the zodiac and the position of the Moon's nodes.

Erasmus King was another science lecturer who designed a novel type of orrery to be made (probably) by George Adams

Fig. 3.31 Thomas Wright's 1740 orrery for 'Ladies and Gentlemen rather than Noblemen or Princes'. (Photograph by the author with permission of the Science Museum/Science & Society Picture Library)

Fig. 3.32 A detail of Thomas Wright's orrery for 'Ladies and Gentlemen. (Photograph by the author with permission of the Science Museum/Science & Society Picture Library)

around 1770, which is displayed in the London Science Museum. The accompanying note reads: "The brass balls represent the Sun, Moon and Earth. Charging the central column caused the supporting rods to be charged. The charge streamed off the small spikes causing the rods to rotate thereby imitating the motion of the Sun, Moon and Earth," Fig. 3.33.

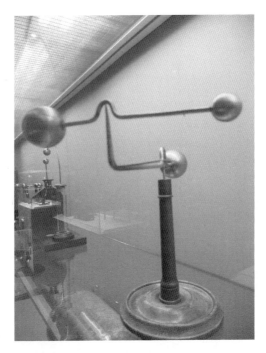

FIG. 3.33 George Adam's electrically driven orrery where the motion is caused by the charge streaming off the small spikes. (Photograph by the author with permission of the Science Museum/Science & Society Picture Library)

There are many other orreries in museums around the world such as a Large Planetarium orrery, London, ca. 1765, by Thomas Heath (1719–1773) and Tycho Wing (eighteenth century). It is owned by All Souls College and displayed in the Oxford Museum of the History of Science. This large planetarium orrery shows all the planets and their attendant moons known at the time. It is driven by a clockwork mechanism regulated by a pendulum situated beneath the base plate. The planets with their respective periods, the Moon, the four satellites of Jupiter and the five satellites of Saturn are fixed to the base plate. The Earth moves along a calendar scale.

Of course, orreries can be found in palaces such as Buckingham Palace, which displays an astronomical clock made by Jean-Antoine Lépine (1720–1814), who was an inventive watch- and clockmaker.

Classic orreries can fetch huge prices. An Ormolu and Mahogany orrery clock made by the famous Raingo Frères in 1830 was sold for around $266,000 in 2009. Hundreds of orreries and orrery clocks have been sold at auction over the years, some for as little as £100 and some for exorbitant prices, such as a boxed Jones portable orrery for £100,000.

Reputedly the finest private collection of clocks and watches in the country is at Belmont near Faversham in Kent, England. It also includes a superb astronomical (orrery) clock. The collection was amassed by George St Vincent, the fifth Baron Harris of Seringapatam, Mysore and Belmont in the County of Kent. The house and gardens are set in a picturesque area of 'The Garden of England'.

Another great port of call with a similar claim to that of Belmont is the Museum of the Worshipful Company of Clockmakers situated at the Guildhall in the City of London. Of course, there are many other such prestigious museums around the world, such as the Patek Philippe in Switzerland, which includes instruments made by the famed Abraham-Louis Breguet.

The practice of clockmaking was considerably advanced by the time it reached America, when wealthy immigrants would bring their clocks with them. Although William Davis, who called himself a clockmaker, hung out his sign for business in Boston in 1683, it is likely that he was more interested in furniture. Clocks that can actually be found, and that can be dated from the 1700s, were made from wood because the brass used was not available overseas. The early clockmakers would make the actual clock, but often had the case made by a skilled cabinet maker. Some instruments are displayed in places where their history is relevant, such as two of the noted Rittenhouse orreries in the United States. One is to be found in the library of the University of Pennsylvania and the other at Peyton Hall of Princeton University. David Rittenhouse (1732–1796) was a renowned mathematician, astronomer and clockmaker. He prepared for a year to observe the transit of Venus in 1769 and was so excited when it happened that he fainted while lying on his back under the eyepiece of the telescope. He recovered quickly and continued his observations. He made his advanced orrery in 1770.

Notable objects are not always to be found in museums and the like. A home environment can be a refreshing place to see them. Sir Patrick Moore kindly allowed the following items to be included. In Fig. 3.34 one of the house cats is assisting to model

FIG. 3.34 A William Jones orrery from around 1794. (Photograph by the author with permission of Sir Patrick Moore)

the tellurium. The name Jones, London, is clearly marked on the Earth globe; it was probably made during the later part of the eighteenth century. It is operated manually by an ivory-covered handle on the side. A somewhat similar model, dated 1794, was sold at auction recently for £8,000. This Jones is the William Jones who formed a distinguished optical company with his younger brother Samuel. Yet another auction house sold a similar one for £3,600 that was quoted to have a label reading: "New Portable Orrery by W. Jones and made and sold by W & S Jones 30 Holborn, London". The National Maritime Museum at Greenwich has a slightly later, but very similar, model, with the 'Barden Family of London' label printed on the Earth globe. One should always check what one is buying; at both auctions, the lot description stated that the model includes the two *inner* planets Mercury and *Neptune*!

Figure 3.35 shows the stunning detail of the print on the hand-colored lithographed paper.

Modern Models and Replicas

Figure 3.36 shows a nice modern presentation model, with the glass protective cover removed, showing the inscription that acknowledges 70 years of Sir Patrick Moore's membership of the British Astronomical Association. Figure 3.37 shows the detail of the inscription. Once more beautifully modeled by Jeanie, Sir Pat-

Fig. 3.35 A closer look at the inscribed detail of the William Jones orrery. (Photograph by the author with permission of Sir Patrick Moore)

rick's other cat, is a modern orrery (Fig. 3.38) assembled at home through the regular subscription for the collection of parts. A detail of the top plate and rocky planets is shown in Fig. 3.39. Copies of some of the eighteenth-century orreries are replicated for sale in amazing beauty and detail, for example Ferguson orreries.

Astronomical Clocks

Clocks with orreries are called astronomical clocks and became very popular. Figure 3.40 shows such an ornate instrument made in mahogany, gold and brass. It was created in 1778 by the British clock- and watchmaker Henry Jenkins (1760–1794) in London, Aldersgate Street, at a time of great interest in astronomy (much accelerated by the work of James Ferguson), the transits of Venus undoubtedly having something to do with that. The upper dial is a heliocentric orrery of six revolving concentric rings painted with gold stars against a dark blue background, each ring having a

138 Orrery

FIG. 3.36 A modern orrery constructed for presentation purposes. (Photograph by the author with permission of Sir Patrick Moore)

FIG. 3.37 Detail for the celebration orrery of Fig. 3.36. (Photograph by the author with permission of Sir Patrick Moore)

FIG. 3.38 A modern design for collecting and building at home. (Photograph by the author with permission of Sir Patrick Moore)

FIG. 3.39 A detail of the home-built orrery. (Photograph by the author with permission of Sir Patrick Moore)

black-and-white sphere to represent one of the six planets known at the time. Much information is contained on the clock, including the durations of the planets' orbits, signs of the zodiac, time, and tide data. The beautiful case was made from mahogany, gold and brass and is to be found in the British Museum, London, UK.

To see the huge number of photographs taken of orreries around the world a great place to include in the search is a visit to internet photo-sharing sites.

Fig. 3.40 An Henry Jenkins astronomical clock, 1778. (Courtesy of © Trustees of the British Museum)

Homemade Models

A search of the internet soon points to many collections of astronomical models and timepieces around the world, and also to sites where one can purchase inexpensive models for kids, such as rotating planet cut-outs hanging from string or wire from the ceiling.

Orrery—the Man and the Model 141

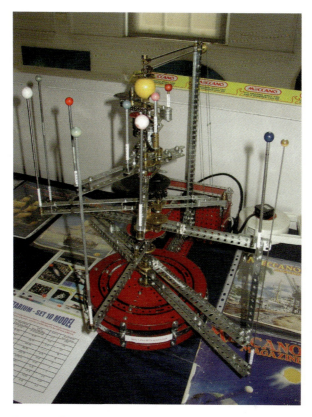

FIG. 3.41 A beautifully constructed meccano model of the Sun, 9 planets (when Pluto was a planet) and Quaroar. (Courtesy of Michael Whiting)

Mechanical constructor-kits have been available for ages, and these are not only for the pleasure of kids! Certainly, simple orreries can be made using meccano, but detailed, printed plans are available to be able to create wonderful models with 9 planets (as designated at the time) and Quaoar (Fig. 3.41).

Lego building-block models can also be assembled, as in Figs. 3.42 and 3.43; a great way for hands-on learning (especially for kids) about the Solar System—as originally intended 300 years ago. A Lego model has even been built of the ancient Antikythera mechanism (SOTT.net).

There are a huge number of virtual working orreries on the internet, many of which are available for download. No images here, of course, as they are movies. Many are from schools and educational establishments. Some are accompanied by horrible (depending on your taste) and loud music. Some take ages to load.

142   Orrery

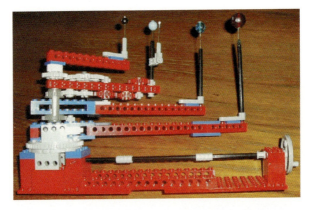

Fig. 3.42 Building-brick Lego model, closed. (Courtesy of NASA and the Kepler mission)

Fig. 3.43 Building-brick Lego model, working. (Courtesy of NASA and the Kepler mission)

Many work well and immediately on some video websites with hundreds of examples, and some are made from wood or metal or bits and pieces. Some are ready-made but assembled through the collection of parts received from regular magazine subscription, see above.

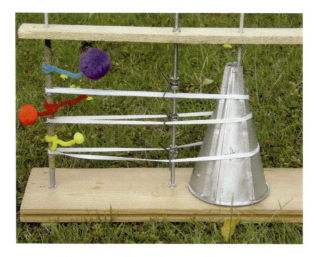

Fig. 3.44 Homemade orrery from bits and pieces. (Made and photographed by the author)

For the more entrepreneurial spirit and a little DIY (Do-It-Yourself) ability, a reasonable model can be made from bits and pieces around the house and purchased cheaply from art and general shops. Better still, especially for teachers, encourage the students at a science club to use their imagination and build one as in the image above. A miscellany of items were assembled to create the model in Fig. 3.44. Adjusting the height of the elastic strips on the cone allowed relative variation of the orbital speeds of the inner planets. A later model incorporated a motor to drive the cone. Great fun!

Human Orreries

Human orreries are great for getting the feel of how the Solar System works. Indeed, they can be much more representative of the actual pathways, though not the angles with the ecliptic, of the planets around the Sun. Teachers can explain the motions of the heavens by placing the children at the position of planets and watch them dashing around to represent the inner planets or creeping slowly for the gas giants. "No one wanted to represent Uranus!" The Armagh website describes its famous model:

FIG. 3.45 Armagh Observatory's Human Orrery. (Courtesy of Armagh Observatory, Northern Ireland)

Armagh Observatory's Human Orrery (Fig. 3.45) is interactive: it allows people to play the part of the moving planets. It features an accurate scale model of the positions and orbits of the Earth and the five other planets in the Solar System known since ancient times (Mercury, Venus, Mars, Jupiter and Saturn), as well as the asteroid Ceres and two comets: 1P/Halley and 2P/Encke. The orbits of these objects are arranged on the ground with stainless steel tiles. Jumping from one tile to the next represents a 16 day time interval for all the planets, except Jupiter and Saturn, whose tile jumps represent a 160 day interval. The tiles for Ceres and the comets have 80 day intervals. More distant objects which could not be accommodated within the dimensions of the Human Orrery are listed on the outer ring of the exhibit. (The observatory also houses an early nineteenth-century orrery by Gilkerson up to Uranus—with moons.)

Solar System Models

The drawback to creating totally realistic orreries is the fact that distances to the planets are so huge that a clockwork scale-model to fit onto the average dining table would be cumbersome at the very least. The smallest planet is Mercury and the largest is Jupiter, so that if the size of Mercury were to be set at 1 in., Jupiter would be 30 in. in diameter and the Sun huge. If the distance from

Fig. 3.46 Otford Solar System in Kent, England, David Thomas and the Sun. (Photograph by the author)

the Sun was set at 1 in. for Mercury, then Neptune would be 77 in. and Pluto even farther away. So it is not an impossible task, but hardly portable. Hence the birth of static-scale models where the appropriate scale distances are marked in some way on the landscape, but are not mobile.

The Sweden Solar System is known as the world's largest permanent scale model in 2013. The Sun is represented by the Ericsson Globe in Stockholm, the largest hemispherical building in the world at 110 m or 361 ft in diameter. It is the national indoor arena for the country. The inner planets are also sited in Stockholm, but the other planets are sited northwards along the Baltic Sea, and the model spans the length of Sweden.

Smaller-scale models are known throughout the United States, Canada, many countries in Europe, in Australia and maybe elsewhere.

Although there are large models that include planets as far out as Pluto, many miles from the 'Sun', the Otford Solar System in Kent, England, is big on ambition and achievement (Figs. 3.46, 3.47 and 3.48). Pluto is a modest distance of around 1,000 m from the 'Sun', sited in the local playing field, but the project also includes markers for the distance to 'stars' in Los Angeles and New Zealand. Unfortunately, the creator of the model, David Thomas (who can be seen in Fig. 3.46), is no longer with us, but he had already been in contact with NASA who were responding favor-

146   Orrery

Fig. 3.47 Otford Solar System in Kent, England, the inner planets. (Photograph by the author)

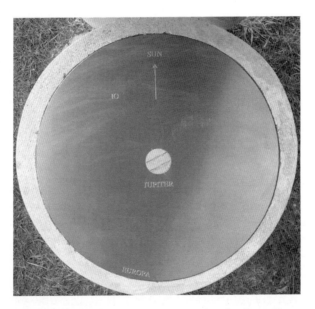

Fig. 3.48 Otford Solar System in Kent, England, the pillar representing Jupiter. (Photograph by the author)

ably to planting one of David's plaques on the Moon in the next available mission.

The following table assists with some of the Solar System data.

Planet	Mean Distance from the sun	Diameter	Mass	Density	No. of satellites (end of 2012, NASA)	Orbital period	Inclination to ecliptic	Inclination of equator	Rotation rate
Mercury	0.39	0.4	0.05	5.1	0	0.24	7.0	?	59 days
Venus	0.72	1.0	0.9	5.3	0	0.61	3.4	23	243 days
Earth	1.00	1.0	1.0	5.52	1	1.00	0	23	1 day
Mars	1.52	0.5	0.11	3.94	2	1.88	1.9	24	24.5 h
Jupiter	5.20	11	318	1.33	50 (+16 provisional)	11.86	1.3	3	10 h
Saturn	9.54	9.54	95	0.69	53 (+9 provisional)	29.46	2.5	27	10 h
Uranus	19.18	4	15	1.56	27	84.01	0.8	98	11 h
Neptune	30.06	4	17	2.27	13	164.79	1.8	29	16 h
Pluto	39.44	0.5?	?0.1	?	5	247.69	17.2	?	6 days

4. A Closer Look at Gear Calculations, Time Corrections, Escapements and Orbital Resonance

Gears

A glance at the more comprehensive orreries that include all planets of the Solar System known at the time and many satellites, whether old or modern, tells us that they are built from a huge number of gears (cogs), pinions (cogged spindles), arbors and platforms each with its calculated angles, tilts and teeth. The complexity and precision engineering is clearly aimed at representing not just the relative positions of the bodies, but the times associated with them; about 365 days for the Earth to complete one circuit around the Sun, around 28 days for the Moon to rotate about the Earth and so on. While ancient civilizations were content to work with such approximations when they were discovered, accurate astronomical predictions demand more, much more.

Toward the end of the seventeenth century, Sir Isaac Newton's natural philosophy, gravity, sparked a new era of widespread scientific interest particularly in the relative movement of bodies. The study of astronomy no longer became the prerogative of a chosen few, but was laid open to the understanding of any literate person of whatever social and educational upbringing. It saw the rapid growth of adult science education, one of the most outstanding and fascinating of all social changes in Hanoverian England. The invention and development of the planetarium or tellurium is a great example. William Stukeley (1687–1765) was a scientist and a friend of Newton and wrote that a planetarium had been made by Richard Cumberland (1631–1718), a rector and later bishop, who was also stated by Stukeley to be an excellent mechanic, philosopher, mathematician and astronomer. Stukeley also referred to an astronomical clock made by Cumberland that he had in his possession and which had "but three wheels and shows hours, minutes and seconds and goes extremely well". Stukeley also made a drawing, preserved in the Bodleian Library, of a planetarium made by Dr Hale that consisted of just three wheels and demonstrated the annual and diurnal motions of the Earth and the orbital motion of the Moon. There was no clockwork drive, nor even a handle, so it must have been pushed round by hand. This was the model that led to an early debate concerning who first built the mechanical tellurium (orrery) designed to illustrate Copernican principles.

The first Graham model, now in the Adler Museum, Chicago, has a clock face with a single hand that sweeps two series of 12 h with quarter intervals, mean solar time. Within the face is a plate that indicates sidereal time in a similar manner. On the top are three engraved circles giving the days of the month, the months of the year and symbols for the 12 signs of the zodiac. The metal Earth is engraved with circles of latitude and longitude. Unlike the second, G&T instrument, this one has a built-in clock drive, spring-driven and with a cord fusee. The 'going period', time between winds, is around eight days. Regulation is by a 10-cm pendulum and the clock movement can be disengaged and coupled to a winding handle. The annual wheel of 16.5 cm has 365 square-form teeth. Seven wheels are designed to rotate the Earth, and the model has 30 wheels altogether, excluding the clock and main drive wheel. Apart from being more ornate, the G&T model at Oxford has the same construction of wheels and teeth without the clockwork, of course. The G&T model indicates the date according to both the Julian and Gregorian calendars. Vayringe, as already mentioned, made a copy (in 1720) of the Graham instrument that has a clockwork drive. That model bears the signature 'Ph: Vayringe Fecit A Luneville'.

Many orreries were made by craftsmen who founded their expertise and trade on the working of wood, which led to the housing of orreries in a usually most ornate box. Even the cogs were beautifully fashioned from carefully selected wood. However, the handsome container hid the majority of the mechanism that was complex, to say the least, and likely discouraged wondering at the

A Closer Look at Gear Calculations, Time Corrections ...   151

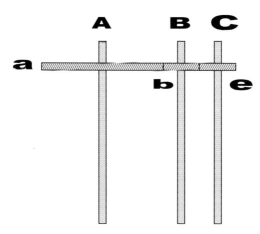

FIG. 4.1. A simple three-spindle mechanism. (Drawing by the author)

masses of wheels and the reasons for the design and interaction of each! The mechanism begins with a drive, a source of energy through clockwork or a manual winding handle or even relying on pushing the planets by hand in a few examples. Appropriate relative rates of angular speed or rotation for each of the model's planets and moons is achieved through the interlocking of cogs between pinions and other cogs. A spindle with the required rotational movement for a particular body protrudes above the others, to which is attached a platform or sphere indicating the body's identity and/or properties. A very simple starter to visualize the mechanism is shown in Fig. 4.1.

As a gross approximation, assume one year has 360 days and each of the twelve months has 30 days. Spindle A needs to be driven one rotation to attach a disk or sphere on top to represent the passing of one year. Spindle B must rotate 12 times faster to represent the months, and spindle C 30 times faster than that. Let speeds be represented by the symbol S and the number of teeth T. The base formula for calculating the resulting rotation speeds or number of teeth is

$$S1 \times T1 = S2 \times T2.$$

This is a statement of conservation: if two gears are in mesh, the product of speed and number of teeth is conserved. For the purposes of illustration, let the number of teeth on cog a for spindle (pinion) A be 360 and the speed of rotation 360 (degrees per rotation or

152   Orrery

per hour or other choice). Pinion B must rotate 12 times as fast to represent the months. This means that

$S1 = 360$
$T1 = 360$
$S2 = 360 \times 12$
T2 (the number of teeth on cog b) we wish to determine.

According to the formula $S1 \times T1 = S2 \times T2$,

$$360 \times 360 = 360 \times 12 \times T2 \text{ or } (360 \times 360)/(360 \times 12) = T2 = 30.$$

So the number of teeth required for cog b on pinion B is 30.

Pinion C is required to go 30 times faster than B and needs a cog e driven by the cog b.

Therefore, $S2 \times T2 = S3 \times T3$
$S2 = 360 \times 12$
$T2 = 30$ (already calculated)
$S3 = 30 \times S2$ (our requirement)
T3 we wish to determine to give our daily rotation.

So, $360 \times 12 \times 30 = 30 \times 360 \times 12 \times T3$

Therefore, $T3 = 1$.

Clearly, we do not want a cog with one tooth! The addition of a second wheel on pinion B will solve that problem, so a guess at an appropriate cog enmeshed on B could be 360 teeth, see Fig. 4.2. Therefore

S2 (the speed of the pinion B) = 360×12
T3 (cog d) we now say is 360
S3 (pinion C) = $360 \times 12 \times 30$ (to give 30 days rotation each month)
T4 (cog e) we want to know.

$$360 \times 12 \times 360 = 360 \times 12 \times 30 \times T4$$
$$T4 = 360 \times 12 \times 360 / 360 \times 12 \times 30 = 12$$

So a cog e with 12 teeth on pinion C is more manageable.

A Closer Look at Gear Calculations, Time Corrections … 153

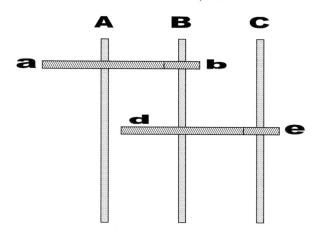

Fig. 4.2 A more sensible structure. (Drawing by the author)

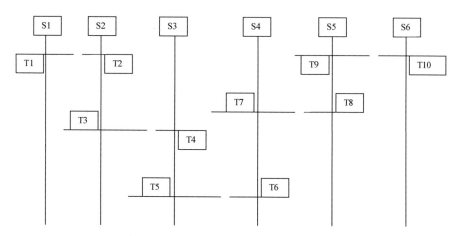

Fig. 4.3 A slightly more demanding set of wheels and spindles. (Drawing by the author)

For enmeshed gears, where one pinion has two cogs keyed onto the middle pinion (ganged), the formula can be simplified since the speed of the middle pinion is the same for both equations.

Note that easy numbers have been used to simplify the maths, and for an insight into the early task of the orrery-makers, one needs to imagine a slightly more demanding set of wheels and spindles (Fig. 4.3) to accomplish the same result of year, month and day with the same approximations of definition but designing more manageable numbers of S and T. Also, stick diagrams are

154 Orrery

easier for quick notes. S1–6 are the speeds of each of 6 pinions. T1–T10 are the teeth numbers of wheels 1–10.

We still want one pinion to represent the month, S3, and another, the day, S6, but to have intermediate gears to make the cog numbers more realistic. S1 is driven but S2 will rotate at one revolution per year, 360°, with a suggested number of teeth on wheel T1 = 60.

S2 required is 1 (revolution).
S1 driven at 60°, 1/6th revolution (say)
T1 60 teeth
T2 = ?

$$S1 \times T1 = S2 \times T2$$

1/6th (60°) × 60 = 1 (360°) × T2
T2 = 1/6 × 60/1 = 10. Therefore we need cog T2 to have 10 teeth.

Reasonable estimates may be made of the specifications of each set of wheels followed by calculations to check. T1–10; 60, 10, 60, 20, 60, 15, 36, 10, 50 and 10. Just in case one should wonder where these figures came from, they are simply suggestions produced from logically likely numbers, the idea being to illustrate the task of the orrery- and clock-maker as an introduction to descriptions of more complex models.

Progressing through the gantry;

$$S2 \times T3 = S3 \times T4$$

S2 = 1 (already calculated)
T3 = 60 (a suggestion)
S3 = to calculate
T4 = 20 (suggestion)

$$S3 = 1 \times 60/20 = 3$$

$$S3 \times T5 = S4 \times T6$$

S4 = 3 × 60/15 = 12 that is, 12 revs and pinion S4 representing the months.

$$S4 \times T7 = S5 \times 10$$

$$S5 = 12 \times 30/10 = 36$$

A Closer Look at Gear Calculations, Time Corrections ...   155

$$S5 \times T9 = S6 \times T10$$
$$36 \times 50 = S6 \times 5$$

S6 = 360 that is, pinion S6 rotates 360 times for one rev of pinion S2; the number of days in a year.

With the appropriate gears to hand, it would have been possible to construct the model to test its validity. However, a clockmaker would find it essential to make as many calculations as possible before entering into the task of using very many gears that would likely have to be made afresh from wood or metal.

The Equation of Time

We know that the lengths of the year, month and day are not precise numbers as above. The commonly accepted length of a year is the sidereal year, Earth's orbital period with reference to the fixed stars. It is 365.25636 days. Taking precession into account, it is 365.24219 days. The synodic day, year, month differs from the sidereal equivalent. The Earth's various precessions, its precise path and speed round the Sun and the influence of the planets all make large or subtle differences to the lengths of days. Of course, this all plays havoc with orrery- and clock- makers, who try their best to provide as accurate a time as is useful. In ancient times of the developing clock the poor accuracy was not a local problem since there was just one clock in the village that everyone used for their appointments. So no one had an excuse for being late!

The length of one day is never the same as the next. Because our orbit around the Sun is not a perfect circle, the ellipticity results in the Earth speeding up or slowing down. In astronomical or maritime terms this is no small and insignificant fraction! It is all very well to demand clocks with an accuracy of a few seconds when traveling huge distances at sea, but what good would that be if it were not known how to correct for daily variations! The irregularity of time was known in ancient times by the Babylonians, and book III of Ptolemy's Almagest was concerned with the 'Sun's anomaly'. The difference between the mean solar time (clock time) and the apparent or true time as measured by a sundial or other instruments for determining when the sun was at its highest, i.e. midday, is known as the equation of time.

As the difference between true and mean time drifted from day to day—then back again—and so on, the exact amounts were tabulated. There are many corrections to be made to attain the accurate time, but some are so tiny that even by today's demanding standards, they can be ignored for most purposes. For instance, the slowing down of the Earth's rotation contributes to 2 ms per day per century. Since, in the Earth's orbital ellipse, there are two slowing and two speeding-up phases through the year, the corrections increase, reverse, pass to negative, then reverse again past zero. Apparent or true time can be ahead by as much as 16 min 33 s, around 3 November, or behind by 14 min 6 s around 12 February with zero adjustments around 15 April, 14 June, 1 September and 25 December. An equation of time table is given in Table 4.1. The earliest table of such an equation of time was published by Christiaan Huygens in 1665.

A graphical representation (Fig. 4.4) illustrates the effects of deviation of the path of the Earth around the Sun from ideal.

There are two major contributions to the curve. The eccentricity of the ellipse, which causes an increase and decrease of speed at locations around the orbit, contributes a sine wave variation with an amplitude of 7.66 min and a period (wavelength) of one year to the equation of time. Second, because the Earth's rotation on its axis is at an angle of 23.5° to the ecliptic, this contributes another sine wave variation with an amplitude of 9.87 min and a period of a half year. The curve in Fig. 4.5 represents the combination of the two sine waves; the axes represent the same times and dates as Fig. 4.4.

As if clockmakers didn't have enough on their plate designing clocks to be ever more accurate! Now they had to display the extra information to cope with this equation of time. These clocks were called equation clocks. They all included a mechanism to simulate this extra adjustment through a lever moving, or a shaft rotating, to effect the necessary movement. In the first type, a shaft is driven by the clock so that it rotates once per year at a constant speed. A cam is incorporated that is shaped in a way that its radius mimics a graph of the annual variation of the equation of time. A lever rests upon the cam and the corresponding movement drives

Table 4.1 An equation of time table. (Assembled by the author)

Date		Minutes	Date		Minutes	Date		Minutes	Date		Minutes
Dec	27	−1	Mar	26	−6	June	27	−3	Sept	28	+9
Jan	1	−3		31	−4	July	2	−4		3	+11
	6	−6	Apr	5	−3		7	−5	Oct	8	+12
	10	−7		10	−1		13	−6		14	+14
	15	−9		15	0		18	−6		19	+15
	20	−11		20	+1		23	−6		24	+16
	25	−12		25	+2		28	−6		29	+16
	30	−13	May	1	+3	Aug	2	−6	Nov	3	+16
Feb	4	−14		6	+3		8	−6		8	+16
	9	−14		11	+4		13	−5		13	+16
	14	−14		16	+4		18	−4		18	+15
	19	−14		21	+3		23	−3		22	+14
	24	−13		26	+3		28	−1		27	+13
Mar	1	−12	June	1	+2	Sep	3	0	Dec	2	+11
	6	−11		6	+1		8	+2		7	+9
	11	−10		11	+1		13	+4		12	+7
	16	−9		16	−1		18	+6		17	+4
	21	0		22	−2		23	+7		22	+2

158 Orrery

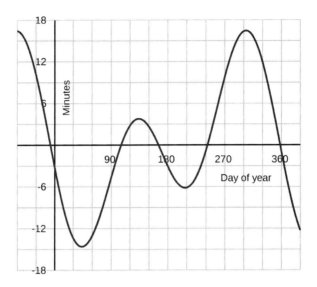

FIG. 4.4 Graphical representation of the equation of time. (Courtesy of Creative Commons Attribution-Share Alike 2.5 Generic, 2.0 Generic and 1.0 Generic license. Attribution; Drini derivative work)

FIG. 4.5 Graph showing the equation of time (*solid line*) along with its two main components plotted separately, the part due to the obliquity of the ecliptic (broken line) and the part due to the Sun's varying apparent speed along the ecliptic due to the eccentricity and ellipticity of the Earth's orbit (*dash-dotted line*). The X-axis is time in minutes and the Y-axis is the time of the year. (Courtesy of Thomas Steiner, Creative Commons Attribution-Share Alike 2.5 Generic license)

other components of the clock. The second type involves a double shaft mechanism such that they simulate the trace of the two-component equation sine waves previously described, one rotating once per year, the other once every six months.

Equation clocks of the time were individual. All were uniquely made and no such clocks were exactly the same. Variously displayed were the different times, mean solar or true (sundial) time, corrected time or just an attached chart of daily corrections to be made to clock time. The cam method, involving the differential gear, was the most versatile. The first was claimed to have been invented by Joseph Williamson in 1720. His clock is shown in Fig. 4.6. The claim was recorded in The Philosophical Transactions of the Royal Society, volume 30, in a letter from the said Mr. Joseph Williamson, Watchmaker, asserting his right to the curious and useful invention of making clocks to keep time with the Sun's Apparent Motion. On a chance reading of a French book he observed that the author of that book claimed to have constructed, invented, a clock to display apparent solar time. Outraged Williamson certainly seems to have been, because he then proceeded to describe a clock found (around 1699) in a cabinet of King Charles II of Spain that had been referred to by a Jesuit, one P. Kresa, in a letter to Mr. Williamson, clockmaker to his Imperial Majesty. The clock showed both equal time and apparent time according to the equation of time. This, Williamson said, was the same as one he had made for Mr. Daniel Quare six years earlier to go to King Charles II of Spain. The mechanism was such that the pendulum was progressively and mechanically adjusted to the Sun's motion. In spite of all this, Tompion may have beaten him to it as a long-case equation clock exists in Buckingham Palace that was made for William III in 1695.

It is also suggested that early clockmakers may not have been "quite up to it" on the mathematics of the differential gear required for compensation for the equation of time and may have used a design by Robert Hooke.

It must be emphasized that the first equation clocks (and hence the equation tables) did not allow for the slow changes, unknown at that time, and therefore are less accurate at the present day, over 300 years later, when the greatest error from the 'slow' causes is about one minute. If the original clocks contin-

Fig. 4.6 Joseph Williamson's Long-case Equation Regulator, around 1710–1720, with cam and differential gear. (Courtesy of the Trustees of the British Museum)

ue to work, they will become increasingly inaccurate. Even allowing for the great strides that the early pioneers made in the advancement in accurate timing it is sobering to know that inexpensive wristwatches are now available that pick up official time and date signals that result in an accuracy of one second in 1,000,000 years!

It is interesting to note that the marvelously robust and accurate clocks of John Harrison were not constructed to indicate the difference between apparent/true solar time and mean solar/clock time, that is, the equation of time. The determination of the exact location at sea was not complete until the accurate time was used in conjunction with the measurement of the true local solar time and latitude.

Measurement of Latitude

Knowledge of the maximum angle between the Sun and the horizon would provide true solar time but would have to be very accurate to avoid making all Harrison's efforts being a "waste of time"! For over two thousand years navigators have known how to determine their latitude in the northern hemisphere by observing the position of the pole star, a circumpolar star that does not 'move'. At the North Pole the pole star is directly overhead or at 90° to the observer on Earth. At the equator the pole star is on the horizon. The 'ancients' had ways of estimating the angle such as holding out a finger or two at arm's length and level with the horizon. An improvement, from about 1200AD, was to use a quadrant, a piece of wood cut to a quarter of a circle, with a plumb bob dangling against a scale of degrees. Viewing the pole star along the upper edge and noting the angle gave the latitude. Before putting out to sea the latitude was noted at the home port. When at sea and out of sight of land all that was required to return to port was to sail north or south until the plumb bob registered the home port latitude, then sail east or west to track the latitude back to port. But the quadrant had its drawbacks at sea with the pitching and rolling of the boat and the wind playing havoc with the dangling weight. One design by 'The Orrery Man', John Rowley, in 1700, is shown in Fig. 4.7. It was a piece of paper to be attached to a board to display not just angles but also much astronomical and mathematical data. A much larger land-based quadrant was constructed in Beijing at the ancient observatory in 1673 (Fig. 4.8). A

162   Orrery

Fig. 4.7 A quadrant designed by John Rowley in 1700, was a piece of paper containing much information to be attached to a board. (Photograph by the author with permission of the Science Museum/Science & Society Picture Library)

Fig. 4.8 Land-based quadrant constructed in Beijing at the ancient observatory in 1673. (Courtesy of Hans A. Rosbach Creative Commons Attribution-Share Alike 3.0 Unported, attribution)

FIG. 4.9 Portable quadrant made by John Bird. (Photograph by the author with permission of the Science Museum/Science & Society Picture Library)

stunning example of a portable quadrant made by John Bird (1709–1776) is shown in Fig. 4.9. A quadrant was one of the instruments taken by an expedition supported by the Royal Society that set out to measure the transit of Venus in 1769. The example may have actually been the one carried by the expedition, but there is some mismatch between quoted dates of production and the date of the transit.

The next major development was made around 1734. It was the octant, which incorporated mirrors, had a shorter 1/8 turn or 45°; the longer quintants or quadrants had 72 or 90°. A mid-eighteenth century octant is shown in Fig. 4.10. It was first developed independently by two men, a mathematician, John Hadley (1682–1744) and a glazier, Thomas Godfrey (1704–1749). It might be that Isaac Newton really invented the reflecting quadrant in 1699 although its recognition depended on Edmund Halley to whom the design was entrusted to publish the description. He didn't, and so the design was only discovered after Halley's death.

The sextant was another step forward. The first instrument was made by John Bird in 1757 and is still being used today. Its purpose was, in principle, to measure the angle between any two objects, which could be the night-sky pole star (northern hemisphere) and the horizon for nocturnal navigation or the Sun and horizon during the day. The instrument in Fig. 4.11 was made about 1775 by Joshua Lover Martin whose father Benjamin Martin

Fig. 4.10 Mid-eighteenth century octant. (Photograph by the author with permission of the Science Museum/Science & Society Picture Library)

Fig. 4.11 Sextant made by Joshua Lover Martin. (Photograph by the author with permission of the Science Museum/Science & Society Picture Library)

Fig. 4.12 A modern sextant

was sometimes credited as the first to supply a microscope fitted with a micrometer in 1740. The name of the sextant arises because its scale has a length of 1/6 turn or 60°. A modern model is shown in Fig. 4.12. The sextant relies on the optical principle that if a ray of light is reflected from two mirrors in succession, the angle between the first and last direction of the ray is twice as large as the angle between the mirrors. This angle can be read off the arc. Figure 4.13 illustrates this.

To produce an accurate altitude of a star, Sun, Moon or other body, a few corrections are vital. These are the index error, dip, refraction, parallax and semi-diameter. More details are not necessary here, but a reference is included later.

Cogs and Orreries

To return to orreries, after laying the foundation for the understanding of wheels, cogs and the equation of time, a good example to inspect is that made by Ferguson. As already mentioned, thorough calculations of cogs and teeth in the wheel-work were carried out before manufacture to produce the most accurate relative

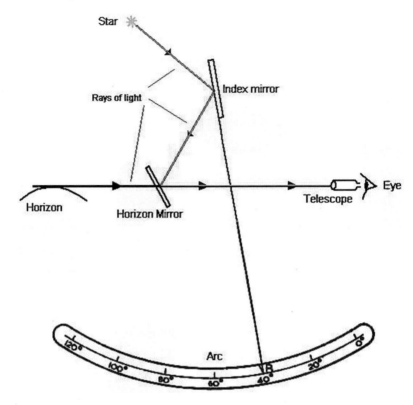

FIG. 4.13 Diagram to illustrate the working of a sextant. (Courtesy of Clipperlight.com)

movements. To achieve this, very high numbers often fell out of the calculations, which were then factorized to describe the specification of wheels to be made. One particular difficulty often arose in that some of the very high numbers were primes and therefore irreducible. A typical example is the following recorded cogitation:

"We may here note that the last lunar ratio but one we here give, 850/25101 cannot be adopted, because 25101 when divided by 9 leave 2789, which is a prime number, and therefore cannot be further reduced; the next ratio above this is 1749/51649, and which can be arranged thus:- 12/52×22/58×53/137 pinion 12 turns round in 24 h and drives a wheel of 52 teeth, having on its axis a wheel of 22 teeth that turns one of 58 teeth, on whose axis is a wheel of 53 teeth, which turns round a wheel of 137 teeth in 29 days, 12 h, 44 min, 2.881 s; therefore, this last train is the most perfect that can be found clear of prime numbers."

A Closer Look at Gear Calculations, Time Corrections ...   167

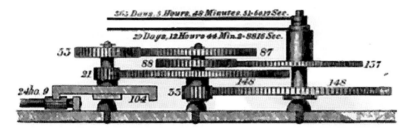

Fig. 4.14 Image from Ferguson's autobiographical memoirs, written in 1756, expanded by Ebenezer Henderson in 1867. The diagram is slightly fuzzy because it is reproduced from an old manuscript. (Reproduced by the author with the permission of Cambridge University Press)

To clarify the legend in the above diagram	
Top bar label	365 days, 5 h, 48 min, 51.6017 s
Bar below	29 days, 12 h, 44 min, 2.8816 s
Drive wheel	24 ho. Pinion 9
Teeth numbers reading from top to bottom	
Left axis	55, 21, 104
Center axis	87, 88, 148, 33
Right axis	137, 148

And, to put all the above into an example of eighteenth century reality, the following diagram (Fig. 4.14) and 'recipe' is reproduced, with the kind permission of Cambridge University Press, from Life of James Ferguson, F.R.S. (2010), edited by Ebenezer Henderson and James Ferguson. (Ferguson's autobiographical memoirs, written in 1756, were expanded by Ebenezer Henderson in 1867.)

"We annexe a sectional plan of the wheel-work, and shall here merely note, that in the section, pinion 9 must turn once round on its axis in 24 hours, drive a contrite wheel of 104 teeth, with its teeth looking downwards, (in order that the last wheel in the train may turn round in the proper direction. Wheel 104 has above it a small wheel or pinion of 21 made fast to it, which turns a wheel of 148 teeth, and has on its axis a small wheel of 33 teeth, that turns wheel 148 once round in 365 days, 5 h, 48 min, 51.6017 s, and a bar on its axis will be carried round along with it in this period. In giving this extremely accurate ratio to the public, we are induced to append to it a train of wheels equally accurate for the period of a mean lunation, which consists of 29 days, 12 h, 44 min, 2.887 s. Immediately above the small wheel 21 (on the axis of wheel 104, of the annual train), is made fast a wheel having 53 teeth, which drives a wheel of 87 teeth, and on the same socket, under it, is riveted a wheel of 88 teeth; these two wheels are here supposed to move round on the shaft of the middle wheel 148, but they may be placed on a stud by themselves in any convenient position; but in the present case, assuming they turn round on the axis of the middle wheel of 148 teeth, it must be understood that they do so, not fixed on said axis, but both turning round together loose upon it, and have therefore no connection with the motion of the said axis of said wheel 148. The last named wheel 88 being fast on the socket of wheel 87, turns round with it in the same time, and turns a wheel of 137 teeth (which turns loose on the axis of the last-mentioned wheel of 148 teeth) in 29 days, 12 h, 44 min, 2.881 s. For wheel 53 of the Moon's train is made fast to the axis of wheel 104 of the annual train, which turns on its axis in 11.5555555555 days:- the lunation, decimally expressed = 29.5305889673 days—therefore, 11.5555555555/29.5305889673 days is the fractional expression of the ratio, and on being reduced according to rule in note 133, it will be found that 4664/11919 is the nearest reducible ratio—11.5555555555 × 11919 = 13773066666660045/4664 = 29.5305889078 days, or 29 days, 12 h, 44 min, 2.88677472 s, the true mean period being 29 days, 12 h, 44 min, 2.887 s, and 4664/11919 = 88/137 × 53/87, or 53/87 × 88/134, the wheels used and as added in the sectional plan. Or the wheel-work may be demonstrated thus: 9/104 × 53/87 × 88/137 = 41976/1239576, and 1239576/41976 = 29.5305889, &c. = 29 days, 12 h, 44 min, 2.881, &c. s., as in former result. Thus we have given to the modern orrery-maker numbers for a mean solar revolution of the earth round the sun, and a mean synodic revolution of the Moon round the Earth, so accurate, that we venture to say that it is not probable they will be surpassed."

Such dedication to accuracy!

Before we leave the cogs and wheels basics, a fundamental property of their interaction needs to be described. This was addressed in a novel way by Ferguson's paradox. It arose at a weekly meeting that Ferguson attended, maybe at a coffee house or tavern, when a watchmaker was demeaning the first person ever to have suggested the three-in-one persons, the Trinity, in the Godhead. When he noticed Ferguson giving him a bit of a stern look, he asked Ferguson's opinion of the trinity. Ferguson responded by suggesting that they talk about the watchmaker's business instead and, rather confrontationally, as to whether the watchmaker understood how one gear wheel turns another? Naturally the response was "of course". Records show how Ferguson continued:

"Then, said I, suppose you make one wheel as thick as other three, and cut teeth in them all, and then put the three thin wheels all loose on one axis, and set the thick wheel to them, so that its teeth may take into those of the three thin ones; now turn the thick wheel round: how must it turn the others? Says he, your question is almost an affront to common sense; for everyone who knows anything of the matter must know that, turn the thick wheel which way you will, all the other three must be turned the contrary way by it. Sir, said I, I believe you think so. Think! says he, it is beyond a thought—it is a demonstration that they must. Sir, said I, I would not have you be too sure, lest you possibly be mistaken; and now what would you say if I should say that, turn the thick wheel whichever way you will, it shall turn one of the thin wheels the same way, the other the contrary way, and the third no way at all. Says he, I would say there was never anything proposed that could be more absurd, as being not only above reason, but contrary thereto. Very well, says I. Now, Sir, is there anything in your ideas more absurd about the received doctrine of the Trinity than in this proposition of mine? There is not, said he; and if I could believe the one, I should believe the other too."

Ferguson then said that he could make such a machine, and would bring it along to show to the assembled company the following week. He did so, and asked the watchmaker to explain it. The watchmaker turned it to and fro, took it to pieces and put it back together again, and confessed that he was thoroughly perplexed. "The thing is not only above all reason, but it is even contrary to all mechanical principles".

For shame, Sir, said I, ask me not how it is, for it is a simpler machine than any clock or watch that you ever made or mended; and if you may be so easily non-plused by so simple a thing in your own way of business, no wonder you should be so about the

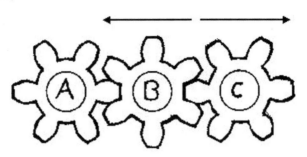

FIG. 4.15 Three cogs in mesh with equal numbers of teeth. (Drawing by the author)

Trinity; but learn from this not for the future to reckon every thing absurd and impossible that you cannot comprehend."

This was rather a naughty way of illustrating a basic method employed in the construction of orreries to produce parallel motion by three equal gears plus a slow advance or retrograde movement. Imagine three cogs with equal numbers of teeth (Fig. 4.15):

Cog A is fixed, and B and C are free to rotate around their own spindle, all attached to a base plate. As B revolves around A in an anticlockwise direction, B naturally rotates in an anticlockwise motion. C then rotates clockwise. Since static A is effectively rotating clockwise relative to B, the rules of contrary motion are conserved. Imagine B and C are on the same spindle, but able to rotate loosely and independently of each other, and mesh with A, i.e. B and C are able to interact with a thicker A. If the number of teeth on B is greater than initially, it would appear to be going backwards relative to A. If C had fewer teeth, it would appear to rotate forward. The effects would best be seen by adding pointers to the extended spindles above. This became known as Ferguson's paradox, and the mechanism formed the basis of his future orrery (about 1750). He progressed from the simple conundrum to the construction of an apparently baffling model. A static wheel was enmeshed with a thick wheel, which meshed with three thin wheels with concentric hollow tube spindles having pointers above. One thin one had the same number of teeth as the thick one, one less and one more. The result of rotation was that as the plate turned on the axis of the static wheel B, and C made apparent movements in the opposite direction. Clearly, there was some fiddling in the first such model because differing cog numbers would not mesh. However, the

A Closer Look at Gear Calculations, Time Corrections ... 171

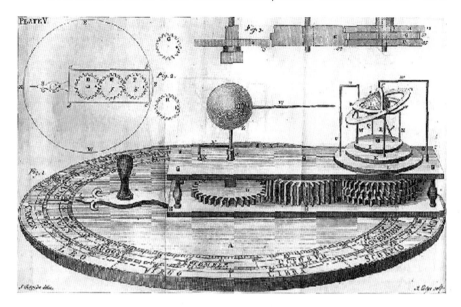

FIG. 4.16 James Ferguson illustration of orrery

original wheels were made of soft wood that allowed the meshing to take place, thus fooling the observer. Any description of the paradox could only really be appreciated by an appropriate movie, many of which are to be found on the internet. An engraving of Ferguson's final orrery (not wood, but appropriately engineered metal wheels) is shown in Fig. 4.16, which might help to illustrate the above.

Escapement

A previous chapter briefly introduced the building blocks of the internal movement of clocks, the weights, springs, pendulum, foliot, verges and so forth. Each improvement made to the construction of one component led to more demands for accuracy and robustness on others, which were consequently improved and highlighted the next weakest link. Eventually, as knowing the time to within seconds over long periods became a necessity, from advances in trading and business, the greatest scientific and precision engineering minds of the age rose to the challenge. One crucial element for the development of ultimate clock precision was the escapement. Since the main player in the history of orreries, George Graham,

172 Orrery

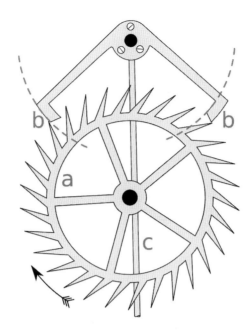

Fig. 4.17 Graham deadbeat escapement. *a* is the toothed (escape) wheel driven by the weight or spring; *c* is the pendulum (foliot replacement) being pushed by the wheel and *b* the pallets. (Courtesy of Frederick J Britten, derivative work, McSush; {PD-US})

was also one of the greatest in the development of the escapement, a little more is said about this mechanism. Bear in mind that we are not talking about modern LCD, LED or digital timepieces, just good, old-fashioned mechanical ones. Quite understandably, to the uninitiated a clock consists of lots of wheels and cogs attached to a spring and fitted into a case with hands and face. Even just flicking through a book dealing with the fundamentals of clock workings or, particularly, one instructing on clock and watch repairs, the complexity is revealed to be awesome!

The escapement is the part of the mechanism that makes the familiar ticking noise of the clock. There are many different designs, such as the cylinder, lever, anchor and duplex and some are known by the names of the inventor, Graham's, Harrison's, Mudge's and so on. The anchor escapement or similar really had to be invented since the verge would require a huge pendulum swing so large that the case would be the width of a cupboard. In the second half of the seventeenth century Hooke rose to the challenge and an anchor design is shown in the diagram in Fig. 4.17.

The arms of the anchor are known as pallets, which have shaped ends (feet, pads or paddles) on the end, b. The shape enables the train, the direct line of energy through the clock mechanism, to be given a kick at each tick, which also causes a momentary recoil, leading to extra wear and tear.

Not unexpectedly, this design is called a recoil escapement. The deadbeat escapement is one in which there is no recoil; no going backwards on the train. The teeth of the escape wheel are shaped precisely to receive, or let go of, paddles on the anchor arms, the regularity of which is controlled by the pendulum, the timekeeping element. The escapement releases the tooth of a gear, which therefore changes from a 'locked' state to a 'drive' state until the opposite arm strikes another tooth on the gear, which locks the gear again. The energy for the mechanism is provided by a spring or weight driving the escape wheel. The flat and correctly aligned/shaped surface of the paddles ensures that there is no recoil. The sliding and pivoting components must be very hard and durable for the clock mechanism to last any length of time and, instead of brass, synthetic rubies are commonly used, hence the description 'jeweled watch'. It is worth dwelling a little more on precisely why the deadbeat was such a significant improvement over the original recoil anchor since so much fame was assigned to Graham. Let's ignore, for the present purposes, the controversy as to exactly who invented the deadbeat. To ensure the clock or watch keeps the correct time, it is essential that a timepiece is isochronous, that is, runs at the same rate regardless of changes in its driving force as the mainspring unwinds. Clocks that use a pendulum are included. Small changes in the friction of the gears or pallets can cause changes in the force driving the escapement and is a major cause of errors. The deadbeat clock is much better than the simple anchor one due to its improved isochronism. In the anchor escapement any changes in the driving force will cause the pendulum to swing faster with minimal increase in the distance of travel of the pendulum bob, the amplitude. Consider two effects. First, the greater force of the escape wheel tooth on the pallet during the recoil part of the cycle tends to decrease the pendulum's swing. Second, the force of the tooth during the forward impulse part of the cycle tends to increase the pendulum's swing. These two effects almost cancel each other out, leaving the amplitude the same, but with a net increase in the speed of swing. So, increasing the swing makes the clock go faster and hence reduces the accuracy. The design solution of separating impulse from locking function in anchor regulators was probably the most important in the entire history of the science and skill of clock-making. It was accomplished in 1675 by Thomas Tompion according to an idea of Richard Townley.

The deadbeat escapement is more immune to changes in the drive force, which results in a better isochronism, although the amplitude of the pendulum's swing is increased. Since the escapement doesn't have a recoil period, increased force causes the pendulum to swing in a wider arc as well as move faster when the tooth's force opposes the direction of the pendulum's motion. The time required to cover the extra distance exactly compensates for the increased speed of the pendulum, leaving the period of the swing unchanged. These comparisons are approximations since there are still tiny inaccuracies because the above opposing forces do not quite cancel each other out. Whatever the movements, the more friction caused by parts rubbing together, the more wear and the shorter the life of the mechanism or the shorter the time between servicing.

A few other significant mechanisms should be noted, although there were so many ideas, possibly 100 following the verge, that didn't make it to the realms of usability.

The detached (detached from the main gear train energy path) detent escapement was invented by Pierre Le Roy in 1748, but it didn't really work. Because it clearly was a good idea, effective models were created by John Arnold in 1775 and were modified by Thomas Earnshaw in 1780. After several modifications by these two clockmakers, the final design of Earnshaw's produced in 1800 prevailed and was used until mechanical chronometers became obsolete in 1970. Detent escapements had many advantages over a lever escapement, one such being that they didn't need oiling, but they were too fragile to be incorporated into pocket watches where the lever mechanism became the norm even though it was a little less accurate. Because the detached mechanism is so important, it is worth stating the principle in a more formal way:

'Dissipation of the oscillator energy imposes the necessity for the regulator impulse function while the regulation of speed itself requires the action of its locking function. If impulse and locking functions are directly supplied by the clock's driving energy, then every change in drive causes variation of impulse, oscillator energy and frequency standard. The invention of the so-called detached escapements resulted from the idea that impulse and locking functions should be completely free from the direct influence of drive, or that the oscillator should be as long as possible free from any influence of the regulator'.

Some of the less familiar terms in the above appear in the glossary, others should be researched should the depth of further interest so warrant.

The cylinder escapement was designed by Thomas Tompion and was patented by himself, Edward Barlow and William Houghton in 1695. George Graham made significant improvements on it in 1722 and it was fitted on all Graham watches after 1726. The center hub of the balance wheel was a hollow tube from which had been cut a portion a little less than a full diameter. As the escape teeth approached (were driven toward) the oscillating hub they were either allowed to pass or were blocked. The noticeable feature of a cylinder escapement is that the wheel is in constant contact with the cylinder and that this contact is frictional. It follows therefore that the escape wheel acts as a real brake on the movement of the balance. This is one of the main flaws of the cylinder escapement. Another major flaw lies in the fact that its swing is necessarily small.

The duplex escapement was another very clever mechanism that employed two sets of teeth on the same escape wheel, alternately interacting with the oscillating balance lever. It was originally invented by Robert Hooke in 1700, improved by Jean Baptiste Dutertre and finalized in a patent by Thomas Tyrer in 1782. At this time, two wheels were involved, but the two separate teeth wheels evolved into one wheel with two rows of teeth later. However, there is constant contact between escape and balance wheels, and contact means friction, resulting in all the consequent disadvantages. It also had tight tolerances and was sensitive to shock, not good for a watch.

Gravity escapement consists of two long levers one either side of a swinging pendulum driven by a weight. They drive a deadbeat arrangement and were essentially used in turret clocks.

FIG. 4.18 Houses of Parliament (Palace of Westminster), London. (Photograph by the author)

Thomas Mudge and Alexander Cumming (1732–1814) designed the earliest in 1766, which was further improved by Captain Henry Kater (1777–1835) in 1830 and others much later. Major problems were finally solved by the British lawyer and horologist Edmund Beckett Denison, first Baron Grimthorpe, Q.C. (1816–1905). The mechanism was called the double three-legged gravity escapement and is incorporated in the clock tower (now known as Queen Elizabeth's Tower, to celebrate her diamond jubilee, 2012) of the Houses of Parliament (Palace of Westminster), London (Fig. 4.18). Clock (Fig. 4.19) plus tower (Fig. 4.20) were finally completed in 1859 by Edward John Dent and, on his death, his stepson Frederick Dent, who was responsible for construction of the clock.

The Grasshopper escapement is famous because it was the invention of John Harrison, who solved the problem of longitude determination and incorporated the mechanism into the first marine chronometer. The mechanism produced great repeatability and also required no lubrication, which was the bane of clockmakers of the age with low-quality and messy lubricants that led to the need for frequent cleaning. It did have its negatives; for example, it was difficult to make. The main reason for its rare occurrence in the following years is that it was invented in 1722, meaning it was closely followed by Graham's deadbeat arrangement, which became widely accepted. A lengthy description would be

A Closer Look at Gear Calculations, Time Corrections ...

Fig. 4.19 Clock of Houses of Parliament (Palace of Westminster), London. (Photograph by the author)

Fig. 4.20 The completed tower with clock of Houses of Parliament. (Photograph by the author)

required to explain the movement of the long legs clicking in and out of the escape wheel teeth and it might still not be clear. However, the mechanism is much easier to understand while watching animations on the internet. A reference is included in the bibliography at the end of this book.

The lever escapement is one of the most important because of its greater efficiency of movement and therefore found in most mechanical watches and many small non-pendulum clocks. Its invention is attributed to Thomas Mudge and it was further developed by others including Breguet and Massey. It is a detached mechanism, which means that the time-keeping element runs separately from the escapement for most of the operating cycle. The concept of 'detached' can be somewhat difficult to visualize but might be helped by this statement: "Whereas, in the anchor (also deadbeat) mechanism the escape wheel 'pushes' the anchor pallets to drive the pendulum, in the lever escapement the lever and balance spring 'allow' the escape wheel to move therefore reducing much friction and energy requirement of the train". The lever, effectively, is a replacement for a pendulum although outside the circuit of the main movement also called the train. On one end of the lever sits a pair of pallets (entrance and exit) lined with a hardwearing ruby that interacts with an escape wheel with specially shaped teeth. At the other end of the lever is a ruby impulse pin on the roller of the balance wheel. A small amount of energy from the escape wheel is transferred to the balance wheel, which is attached to a hairspring that allows the lever to alternate the motion of the pallets for advancement and timekeeping. Figure 4.21 shows the escape components. A full and followable description of the movement is not suitable here and, once again, the best way is to watch animations to be found on the internet, where it then appears quite simple.

Figure 4.22 shows a Thomas Mudge clock from 1768 preserved in the British Museum with the following description: "Gilt-brass and tortoise-shell cased travelling clock with lever escapement, quarter striking and quarter repeat".

Following his experimental clock of c.1754 in which he incorporated his first lever escapement Thomas Mudge went on to make only a small number of timepieces incorporating the new escapement. The best-known of these, the watch made for Queen

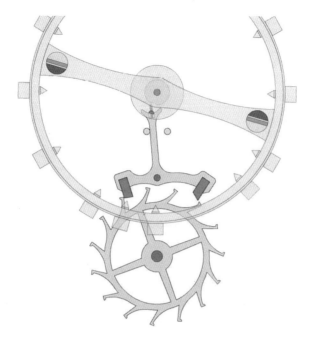

Fig. 4.21 Swiss lever escapement. (Courtesy of Creative Commons Attribution-Share Alike 3.0 Unported, attribution; Shannon)

Fig. 4.22 A highly complex travelling clock, made around 1770, is thought to be the second of Thomas Mudge's clocks to contain his new lever escapement. (Courtesy of the Trustees of the British Museum)

Charlotte in 1769, is still in the Royal Collections. This little traveling clock is thought to be the second of Mudge's clocks to contain his new lever escapement. It is without doubt a tour de force of the clockmaker's art and typical of his work.

This marvelous eight-day clock strikes the hours and sounds the last hour and quarter at the pull of a cord. The movement is a highly complex arrangement of gear trains and striking mechanisms, all arranged in a full-plate construction, making it an extremely demanding piece to work with. The lever escapement is mounted on a platform at the top of the clock and is displayed beneath a round glazed panel in the top of the case. The hours and quarters are struck on two bells mounted on the back plate.

The dial is of the highest quality in white enamel with Arabic minutes, Roman hours and a subsidiary dial at XII o'clock for seconds, numbered every fifth second 5–60. The gold hands for hours and minutes are very finely finished and the seconds are shown by a slender blued-steel hand. In the lower center the maker's name is simply inscribed in capitals, THO MUDGE LONDON. The movement and dial are housed in an unusual cylindrical tortoise-shell-veneered case with gilt-brass bezel and ball feet, and there is also a mahogany carrying-case for protection during transport.

The inner workings of the clock are shown in Fig. 4.23 where the fusée and spring-containing barrel can be seen as can the chain from the fusée to the barrel.

Important inventions are still being made. A modification of the lever escapement was made in 1974 and patented in 1980 by George Daniels, an English clockmaker. The invention makes lubrication of the pallets unnecessary and is considered by many to be one of the most significant advances since that of the lever escapement.

What can an Orrery Illustrate?

Apart from being a great idea and an opportunity to showcase a 'Fecit' talent, what are the features of an orrery that justify its existence and all the fuss? At the beginning of the eighteenth century, the prize for the first one was the accolade of the invention of a new gizmo, especially since it was produced to the quality of one of the greatest celebrities amongst the clockmaking and scientific community. Its production certainly underlined the progressive interests of Graham and, to some extent, Tompion, from clocks and watches to astronomy and the geophysical relationship of the

Fig. 4.23 The inner workings of the Mudge clock, where the fusée, spring-containing barrel and chain from the fusée to the barrel can be seen. (Courtesy of the Trustees of the British Museum)

Earth to other bodies in the cosmos. The model was so novel and the presentation so exquisite that the noble, rich and famous, not to mention the royals, just had to have one for their collections, which is reflected in the fortunate number of preserved early instruments with which to stitch together its history.

As the dust of first revelation settled, the model's application to the illustration of the Earth's position in our local existence in the universe became apparent. It gave what some referred to at the time as "God's view of our Solar System" and certainly underlined the Copernican heliocentric description of its structure. The dynamics of the Solar System and its interacting bodies are characterized by many parameters, so many that it is a wonder it is so stable! In fact it is (thankfully) on a human scale of longevity, although on the longer term some instabilities could become critical and hurl moons and planets out of their comfort zones to cause mayhem. After all, a rock the size of Mars labeled Theia by astronomers was thought to have crashed into the young Earth (the big whack!) to form the Moon from the debris. A brief visit to some of those parameters might be useful to discuss where the orrery could or could not be inserted into astronomical education in addition to sitting on a shelf looking pretty.

Some common terms that crop up are axis, ecliptic, nodes, libration, resonance, orbits, N-body problem, apsides and precession.

Everyone will have noticed that our Moon seems to appear unpredictably all over the sky, sometimes almost overhead, sometimes only just visible above the horizon. But it appears regularly, so it must have a simple orbit easy to describe. Certainly not! And the Sun is at an angle that has its own effects on the system. Gravity has a force of attraction over such unimaginable distances that the planets, Sun, Moon and all the other rocks, condensed masses of gas and debris are sensing each other's positions and movements. Best to ignore the effects between galaxies and the rest of the content of the universe for now, but it has long been known that due to gravitational attraction, the galaxy of Andromeda (M31) at 2.4 million light years away is heading toward us—but will it just miss, hit with a glancing blow or plough directly into us? Using the Hubble Space Telescope, astronomers have calculated the most likely possibility—and it is a head-on collision! The trillion stars of M31 will reach us in about 3–4 billion years time at its present closing speed of about 400,000 km/h—then we will know! However, there are so many other events that will end life as we know it well before then, so why worry. Also there is the imminent (a billion years later after the Andromeda incident.) demise of our Solar System resulting from the explosion of our own star.

The most fundamental facet of the Solar System is the ecliptic, the plane or plate on which the Sun and all the planets and moons sit. For it to be like a flat plate makes sense since it was formed from the condensation of gases and particles revolving furiously like a spinning top (accretion disk) and resolving themselves into lumps of rocks or gases parking in mutually acceptable slots around the massive power house, the Sun. But the Sun is not just a hot blob, it spins with a period of rotation about 29 days

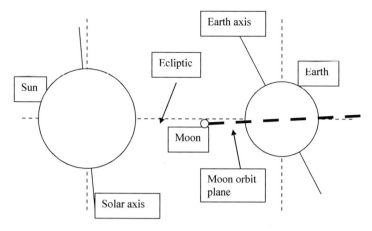

Fig. 4.24 A summary of the relative orientations of the Sun/Earth/Moon system. (Drawing by the author)

synodic (relative to the Earth) and around 25 days sidereal (relative to the stars). These are average rates of spinning since different latitudes revolve at slightly different rates. It is, after all, a massive ball of gas, not like solid rocks that have just the one rate of spinning. Because it spins, it has an axis, which lies at an angle of 7° to the ecliptic.

The simplest orreries contain the Sun, Earth and our Moon. The spinning Earth sits on the ecliptic with its axis at an angle of about 23°. The Moon has its orbit around the Earth with an inclination to the ecliptic of about 5° and a sidereal orbital period of 27.3 days, which is the same as its axial rotation, hence the Moon keeps the same face toward the Earth. It is gravitationally captured. A summary of the orientations are shown in the following diagram (Fig. 4.24). Obviously it is not to scale otherwise, relative to the Sun, the size of the Earth would be represented by the size of a full stop on this page a couple of broomstick lengths away.

To follow the cycle of these elements of a tellurian (Earth/Sun/Moon model), the plane of the Moon's orbit and the orientation of the Earth's axis must be moved on the ecliptic around the Sun, with their accompanying periodicities. Whilst that might seem complicated, it is relatively simply demonstrated with an orrery. The term 'orrery' will be adopted to cover tellurian, lunarium and planetarium because is commonly used. The complete

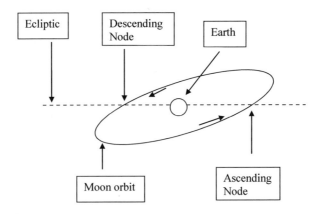

FIG. 4.25 Nodes of the Moon orbit. (Drawing by the author)

cycle, the nodal period when the Earth and Moon come back to the starting point, takes 18.61 years but is not the same as the Saros cycle (a name first given by Edmund Halley in 1691), 18.03 years, which is when all three bodies are in very nearly the same position in the sky for lunar eclipses. The Saros cycle is significant for the prediction of those two eclipses but, of course, there are other eclipses between, but not with the bodies being in the same place.

Nodes are a most important aspect of the movement of bodies. They are the two points where an orbit intersects another orbit or a plane such as the ecliptic. Figure 4.25 illustrates this. The line joining these points is known as the line of nodes. These are not commonly built into orreries except where the orbit of the Moon around the Earth is constructed.

All these descriptions lead to the fact that they might not be too easy to understand and visualize but an appropriate orrery would do the job perfectly and here is where the most important skill of an orrery-maker comes in. All movements must be engineered precisely (and accurately). The more precise (and accurate) the engineering the more turns can be made to observe the longer-term features, which is why the best orrery-makers were the best clockmakers. But how far do you go in building an orrery? Rotation of the Earth would assist learning about night and day. A correct angle of the axis would help with explaining the seasons. A correct rotation of the Moon with one side painted black would illustrate the constant observation of just one side of the Moon facing Earth.

A Closer Look at Gear Calculations, Time Corrections ... 185

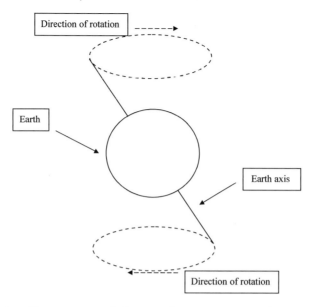

Fig. 4.26 An illustration of the Earth's precession. (Drawing by the author)

Precession is the name given to the rotation of, say, the Earth's axis about its central point (Fig. 4.26) and is important to the long-term understanding of weather. The angle of the Earth's tilt itself varies by about 2.4°. Precession has a periodicity of around 26,000 years and that of the tilt is 41,000 years. It is very important to note that orbits of celestial bodies are rarely (possibly never) perfect circles. They are ellipses and the ellipse of the Earth's orbit rotates relative to the ecliptic with a periodicity of around 25,000 years. These latter three factors are incorporated into what has become known as the Milankovitch cycle, which is used in the prediction of long-term trends in the future weather and, indeed, the survival of mankind. Clearly, it would be too much to expect of an orrery, or orrery operator, to have anything to do with these periods of enormous length. The Moon also wobbles in many ways, but most noticeably in longitude and latitude. Because of the wobble, libration, 59 % of the near face of the Moon can be seen, but not all at once. Again, this does not concern the orrery.

There were some accusations that an orrery does not properly conform to the Copernican system, and it doesn't, because the orrery only enables circular orbits. But that would be a bit fussy. It very much depends on what is expected of the model. In the early eighteenth century not everyone was aware of the movements of the Moon and, as the model became more available to teachers, it gave a greater start to the understanding of our Solar System. Maybe cams could have been introduced to mimic elliptical behavior but there would not be the compulsion to deviate from a perfect circle as it was for the design of a cam by Hooke to progress movements to present an accurate equation clock. It is mind-boggling to know that the Antikythera mechanism, over two thousand years ago, had an additional sliding pin and eccentric small wheel to mimic the ellipticity of orbits that finally explained the purpose of a larger wheel and its number of teeth.

With the addition of planets came more understanding by the pupil but more of a realization that the model was just indicative of how the Solar System worked. Scale had to be sacrificed. To appreciate what the model had to offer, a study of some individual orreries might assist. The addition of planets to the orrery highlights another important aspect of celestial mechanics: resonance. The Solar System can appear to be quite manageably small compared to the universe so that it is easy to visualize the placing of planets and the interaction of their forces as if on a table top. Each planet has its own character and 'the table top' is filled with zillions of rocks of various sizes. Most of the planets have their moons and the regular visitors from outer space, comets, plough through the lot of them on their way to and from the Sun. How could the positions and movements of that mixture be other than random or chaotic? As so often happens, gravity sorts it out.

To begin with the Titus-Bode rule, in its most simply quoted form, states that with increasing distance from the Sun, each larger object (e.g. planets and dwarf planets) is located at about twice the distance of the previous one. A swift calculation reveals that the ratios are not too good. In fact, the rule uses ratios of the semimajor axes of the orbits and relates them to Earth having a value of 10. The theory was quite innovative for the date of first proposal of the observation by David Gregory, 1715, especially as only planets out to Saturn were known. But there were difficulties with gaps in the progression until the discovery of Uranus and, as predicted by the rule, Ceres, the largest body in the asteroid belt. Pluto only fits with the elimination of Neptune. The rule was eventually discredited due to the wealth of modern astronomical data available. Nevertheless, the rule is part of the history of strivings to understand the Solar System. Strange as it may seem the orbit of one body can, however, influence the orbit of another. They constantly pull and tug at each other causing a resonance until either one is kicked into another orbit, or is flung out of the system entirely to remove the resonance and is exiled forever, or they settle into a mutually acceptable configuration where resonance maintains the stability. A stable resonant relationship greatly enhances the gravitational interaction between the two. The ratios of interacting component values usually comprise small integers. A simple example is that of Pluto and Neptune. Since the sidereal orbit of Pluto is around 248 years and Neptune about 164 the resonance is a ratio of 3:2. An orrery that includes these later discovered bodies could well cope with revolving Neptune three times round the Sun while Pluto revolves twice. A little more of an orrery challenge would be the moons of Jupiter, Io, Europa and Ganymede, which have a 1:2:4 resonance. The effect of resonance pervades the whole of the Solar System and is responsible for the marshalling of everything having a mass into its presently observed position. Saturn's moons Tethys and Mimas are in resonance 2:4, Dione and Enceladus 1:2 and Hyperion, Titan 3:4. Many more examples involve the much smaller bodies such as asteroids and Kuiper belt objects. In the history of our Solar System, the late heavy bombardment is a period of time approximately 4.1–3.8 billion years ago during which a large number of impact craters were formed on the Moon, Earth, Mercury, Venus and Mars. It is postulated that the Solar System became catastrophically unstable when Jupiter and Saturn reached a 2:1 orbital resonance, causing the outer Solar System to be unstable and fling Kuiper belt objects and asteroids all over the place to cause impact events on other bodies. There are also other theories.

The interaction of the orbits of bodies leads to 'clearing of the neighborhood' that refers to a (usually large) body using its mass dominance to selfishly clear smaller ones from its orbit. In some instances it cannot quite clear them all out so causes them to be stable in particular relative positions of its orbit. Those that don't conform are slung out into other orbits or elsewhere, maybe even into outer space, but some might be taken into the care of the planet and become a satellite. Those groups that remain in a stable part of the planet's orbit are known as Trojans and two very large such groups are known to reside in the orbit of Jupiter. Even so, the orbit, according to definition, has been cleared and this definition has been used to define a planet. It was decided in 2006 by the International Astronomical Union (IAU) that a planet in the Solar System would be;

1. in orbit around the Sun,
2. have sufficient mass to assume a nearly round shape (in hydrostatic equilibrium),
3. has cleared the neighborhood around its orbit.

The bad news for Pluto was its downgrading to a dwarf planet because it failed on the last of the three requirements. However, it was good news for some of the asteroids or Kuiper belt objects such as Ceres, Haumea, Makekake and Eris, which were promoted to that title. After the initial flurry of discovery of large lumps, dwarf planets, in the Kuiper belt, but now no more, it may be that there aren't any more or that they are at least very scarce. Further out, about half way to the nearest star, there is the Oort Cloud, a huge and wide belt possibly harboring a myriad lumps of rock or ice, and who knows what might lurk there! It is very important to add that there were many dissenters with regard to these definitions of a planet and the topic will surely be revisited at some time in the future.

There are times, very exciting ones, when the orrery should be left on the table and one should go outside to observe at first hand (eye) what is going on. First of all, the Moon. It might be asking a bit too much to calculate the position of the Moon, then observe it to see how correct were the calculations, but just look at that glistening jewel with its familiar patterns of light and dark. Constantly noticing the familiar pattern (Fig. 4.27) and no other, confirms, of course, that the same face is towards us thus illustrating what the orrery illustrates: the axial rotation of the Moon is the same as its orbital period around the Earth.

Comparing the movement of the model with the actualities in the sky is wonderful proof of a theory to a newcomer to astronomy. The bright side of the Moon, of course, always faces the Sun, but from the Earth it appears as a thin crescent sliver soon after

FIG. 4.27 Face of the Moon that always points to the Earth. (Photograph by the author)

New Moon, increases to a full disk and recedes to another thin crescent sliver before 'disappearing'. These ever-changing phases can be amply and simply demonstrated using the orrery, thus supporting the purpose of the early days of the model as bringing education to the masses through operation by teachers and philosophers to their pupils.

The explanation for the seasons can be scientifically demonstrated with the model by pointing to the tilt of the Earth's axis from the ecliptic and how it relates to our path around the Sun.

Actual observations for the beginner astronomer get a little more complicated, but not impossible, when dealing with the planets; the easiest is Jupiter. There are, potentially, a great many moons or satellites for Jupiter, but they all have to earn their place in the system by being approved by the IAU. This entails proving their existence by independent verification and defining their parameters such as the orbit—otherwise, they may never be spotted again. Then new names are proposed and considered for a while before being fully adopted. NASA states: "as of 31 December 2012, Jupiter has 50 officially named moons and 16 more unofficial ones still under consideration". Each planet has its own theme when

Fig. 4.28 Portrait of Galileo. Print made by Ottavio Leoni in 1624. (Courtesy of the Trustees of the British Museum)

naming moons (and craters and other features for the rocky planets), and Jupiter's moons are characters in the life of Zeus or Jupiter in Greco-Roman mythology. Some names proposed in 2009 for new Jovian moons are Themisto, Iocaste, Harpalyke, Praxidike, Taygete, Chaldene, Kalyke, Callirrhoe, Megaclite, Isonoe, and Erinsome.

Until 1610 no one was aware of any moons associated with Jupiter until Galileo Galilei (Fig. 4.28) improved upon an optical invention, a spyglass, which was, after much wrangling, credited to Hans Lippershay (1570–1619), a German-Dutch spectacle maker. He saw four 'stars' circulating Jupiter and, by plotting their movements, correctly identified them as moons. These four moons were thence, not surprisingly, known as the Galilean moons and were observed through even primitive telescopes because they are the largest and brightest. They are now known as Ganymede (the largest moon in the Solar System), Callisto, Europa and Io. Galileo

Fig. 4.29 A composite creation of the moons and Jupiter's great red spot. (Courtesy of NASA)

would have been astounded if he could see what they look like through today's massive telescopes. Figure 4.29 is a composite creation of the moons and Jupiter's great red spot. It has been cited as possible that Robert Hooke was the first to discover this. A mission to the moons of Jupiter, JUICE (Jupiter Icy Moons Explorer), is due to launch in 2022 with the purpose of investigating in more detail hints of habitable environments on the three largest Galilean moons gathered by NASA's Galileo mission in the 1990s. It is strongly suspected that they have liquid water beneath the surface and where there is water there could be life, even though it is highly likely to be very primitive.

So much for the professionals but what can be seen by the beginner amateur? With access to a reasonably good pair of binoculars, say, 8×25, the moons can be seen very close to the planet, although keeping the binoculars steady is a must, maybe on a tripod. Make a sketch of the positions and repeat as many nights as possible. The five bright dots (that includes Jupiter) will appear in a line and will move along the line, changing relative positions

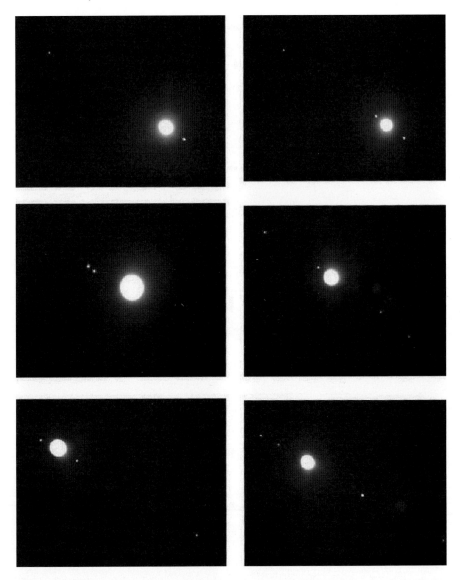

Fig. 4.30 The constantly varying positions of the Galilean moons around Jupiter. (Composite photographs by the author)

and even disappearing when they go behind or in front of the planet. If it is possible, then pop along to one of the many astronomical societies and they will be pleased to show you a much larger version (Fig. 4.30). The positions can be compared with one of the more comprehensive orreries for demonstrations.

The discovery of the moons of Jupiter had a vital part to play in recognizing possibly the most important fundamental unit in the universe, not just the speed of light, but the fact that it had a speed at all! Until the middle of the seventeenth century, the transmission of light was, unsurprisingly, accepted as being instantaneous. Although the longitude problem was solved, as described earlier, with Harrison's chronometer in the mid-eighteenth century, Philip III of Spain (1578–1621) also recognized its necessity for success at sea and out of sight of land and offered a prize for its deterination. The brilliant Galileo used the Jovian system as a cosmic clock based on the eclipses of the moons and presented it to Philip III in 1616. However, the method had many of the usual difficulties such as the constant inability to observe Jupiter (the weather) and the motions at sea. The Danish astronomer Ole Christensen Rømer along with Jean-Félix Picard, a French astronomer, observed a huge number of eclipses of Jupiter's moon Io from Uranienborg, as did Giovanni Domenico Cassini (1675–1712) from Paris. Through the comparison of the timings the longitude difference between the two locations was calculated. Cassini almost casually suggested the possibility that light had a finite speed in 1676. Rømer then went to work with Cassini and concluded that the times between eclipses got shorter as Earth approached Jupiter. Rømer didn't take this further to suggest a speed of light but Christiaan Huygens made the first attempt at an estimate from Rømer's figures to suggest 16 2/3 times the diameter of the Earth per second. Not a bad estimate, but it was not until 1809 that Jean Baptiste Joseph Delambre (1749–1822) got very close to the correct speed of around 300,000 km/s using observations on the Sun.

The orrery came into its own (although with limitations) during 2004 to assist with explaining the astronomically spectacular event of the transit of Venus. Venus crosses the face of the Sun, as seen from Earth, with a precise periodicity of about 243 years that comprises pairs of transits eight years apart and separated by long gaps of 121.5 and 105.5 years. The regular pattern is a result of resonance, as described above, with the ratio of 8:13 for the orbital periods of Earth and Venus. There is no doubt that ancient civilizations were aware of Venus in its two guises, the morning 'star' (Greek, Phosphorus) and the evening 'star' (Greek, Hesperus).

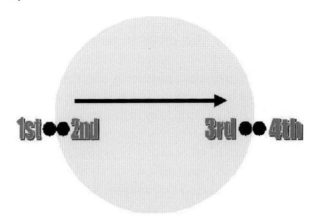

FIG. 4.31 The four recognized points of contact that define a transit. (Drawing by the author)

That is not surprising since, at its brightest, it is the brightest object in the sky other than the Sun and Moon. Predictions of a transit had been made earlier, but the first observation was made by Jeremiah Horrocks (1618–1641), an English astronomer who was the only one to accurately recalculate Kepler's predictions and realize that the pairs of transits were eight years apart. It must have been nerve-rackingly exciting when the clouds parted one afternoon in 1639 to reveal the black spot against a bright Sun. William Crabtree (1610–1644), also an English astronomer and friend of Horrocks, observed the event. Captain James Cook was sent to observe the transit in Tahiti in 1769 to obtain an accurate measurement with which to calculate the distance from Earth to Sun, as suggested by Horrocks.

Four points of observation are accepted as important for a Venus transit (Fig. 4.31): first contact; the point at which the spot fully enters the solar disk; the point where the black dot just touches the other side; and the last point is when the black disk of Venus leaves the bright disk of the Sun. The photograph shown in Fig. 4.32 of the very black dot of Venus against a brilliantly bright Sun disk was taken from the garden of Sir Patrick Moore (Fig. 4.33) on a clear, sunny day on June 8, 2004. Although a manual orrery has limitations in illustrating the transit computer-generated ones would cope admirably.

While Venus transits can be observed with the naked eye (with appropriate filters, of course, for safety) transits of the other

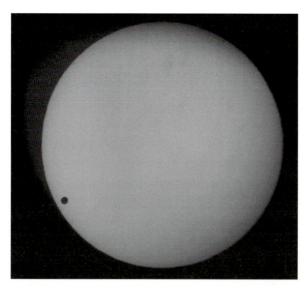

Fig. 4.32 The black dot of Venus transits across a bright Sun disk. (Photograph by the author)

Fig. 4.33 Observing the transit from the garden of Sir Patrick Moore. (Photograph by Chris Buick)

inner planet, Mercury, cannot because it is much smaller and farther away. However, they do occur much more frequently, the last three being 1999, 2003 and 2006. The next one will be in 2016.

So, orreries can be used in conjunction with observations to bring knowledge and experiment together—learning.

FIG. 4.34 A detail of a W&S Jones orrery. (Photograph by the author with the permission of Sir Patrick Moore.)

Studying the 1794 Jones orrery a little more closely was very revealing; we call this the SPM model. It has the expected print of "Design for the NEW PORTABLE ORRERIES by W. JONES" and a second panel "and Made and Sold by W. & S. Jones, 30 Holborn LONDON" (Fig. 4.34).

It has not been restored and is probably in its original condition (Fig. 4.35) and prone to backlash and it even has a bent nail as a split pin in one of its spacers. The Moon phases indication plate and other parts could do with some attention. Of particular note is the small sphere representing the Moon with some blotches on one side to represent (presumably) the cratered surface and the other half bright to follow the movements on turning the ivory handle and illustrate gravitational capture. Mercury is represented by a small white sphere and Venus by a golden ball. Maybe its lack of renovation, maintaining the original, is its value! Compare this with the almost identical portable orrery in the National Maritime Museum, Greenwich, museum ID, AST 1062 (Fig. 4.36) which was also made by S&W. Jones and is dated 1794. The Earth globe map design is the same and both of these are the same as that in another portable orrery (tellurium, National Maritime Museum, Greenwich, AST 1059) made by the Barden family a few years later, around 1800. The SPM Moon sphere is positioned correctly with the 'dark side' facing away from the Earth. In AST 1062 the Moon is now a small, uniformly white ivory ball with no differentiation between the 'light' and 'dark' side, but

A Closer Look at Gear Calculations, Time Corrections ...   197

Fig. 4.35 A full view of the W&S Jones orrery. (Photograph by the author with the permission of Sir Patrick Moore)

Fig. 4.36 For comparison, a similar portable orrery also made by S&W. Jones and dated 1794. (Courtesy of the National Maritime Museum, Greenwich, UK)

very similar to Mercury in the SPM. The AST 1062 look-alike Moon now appears as a larger sphere, but in the position of Venus. A more appropriate graphic on Venus might be something red and fiery. Interestingly, the Barden tellurium, AST 1059, has a small ivory white sphere for its Moon. An explanation for the apparent anomalies will be sought.

5. The Clockmaker's London

London

Charles Boyle, George Graham, Thomas Tompion and John Rowley were all vital to the emergence and evolution of the orrery. A host of other eminent scientists, astronomers, politicians, noblemen, writers and war leaders were also key to the story, but has history been fair to this supporting cast? Two areas of London, Fleet Street and Greenwich, were ubiquitous in their appearance during the research of the fascinating tale and trail of the orrery.

Fleet Street

The history of the ancient city of London is vast, but even selecting the history of a single street reveals a seemingly never-ending quantity of facts, stories, events, anecdotes and biographies. Although there is archaeological evidence for habitation of sorts in London as early as 3000 BC it is likely that it was the Romans who began the first settlement, around 300 acres, which grew to be a significant city a few years after the invasion of AD 43. Queen Boudica knocked it all down and the Romans built it up again to a quality that enabled London, with its large buildings and 60,000 dwellers, to take over from Colchester as the capital of Britannica. The city flourished and a defensive wall was constructed that included several gates, for example Ludgate, Aldgate, Aldersgate, Cripplegate, Bishopsgate and Newgate. Moorgate came much later. The black line in Fig. 5.1, which shows London in 1300 AD, denotes the position of the wall. It can also be seen that Fleet Street and Whitefriars were very much in the countryside in 1300. After the fall of the Roman Empire in 410, the occupation of this country came to an end and London rapidly fell into decline. The rise of the Anglo-Saxons caused the rebuilding of London which, over the centuries, was fought over by the indigenous population

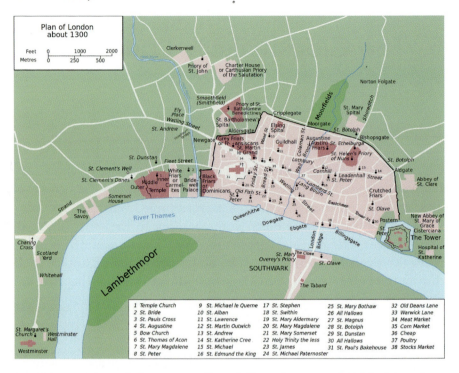

Fig 5.1 Position of the city wall and Fleet Street and Whitefriars in a 1300 AD map. (Courtesy of Creative Commons Attribution-Share Alike 3.0 Unported, attribution; Grandiose)

and the Vikings. Fast-forward, then, to 1066, when the Normans invaded over a spat about who was the heir to the throne and conquered the whole of the land.

By the year 1300 the city's population had exploded to 80,000. There was, however, a serious paucity of services. Waste food, dead animals, horse's droppings and even human faeces were left or chucked onto the streets. Excrement of all sorts was thrown out of the window without consideration of the consequences. And why not? There were no toilets! London has a great many rivers and about 20 of them flow underground, although that has not always been so for some. The largest is the River Fleet and from its headwaters at Hampstead Heath it flows 6.4 km to join the River Thames underneath Blackfriars Bridge. Its derivation is probably from the Anglo-Saxon name for a tidal inlet, 'fleot', when the river served as a dock for ships. Boudica was said to have fought an important battle against the Romans over the

Fleet, hence the old name for Kings Cross, under which flows the river, was Battle Bridge. Possibly the first record of the existence of Fleet Street is in 1274. The street was more extensive than today and included Ludgate Hill and The Strand. As London and its industry developed and expanded the solution to the poor state of the roads and tracks was to shovel the mess into the river, which gradually became a sewer. As time went on the normal rush of water was overcome by the amount of waste, and blockages occurred. The environment may have been awful for humans but was great for rats, which came in from abroad on the busy boats and bred and spread rapidly throughout the wooden houses and dark narrow streets. There was, however, another unseen passenger shipped in unbeknown to the transporters, the Black Death, Bubonic Plague. The Yersinia pestis bacterium had been rising on the continent, probably hopping its way from China or the Gobi desert, but inevitably finding its way to England. In 1348 it was estimated that half the population was killed and bodies just fell in the street where they were gathered up and loaded into carts for disposal (Fig. 5.2). Efforts were made to get rid of the waste, such as using it to build walls/mounds or jetties out into the Thames, but to no avail. The disease disappeared of its own accord and it took 150 years to recover the population. But lessons had not been learned, or, to be more generous, the cause of the illness was unknown and the plague returned with vengeance in 1665. The next disaster, the Great Fire of London, that incinerated a huge proportion of the houses, had the beneficial effect of severely reducing the rat population and certainly must have sterilized some areas. By that time the age of the Enlightenment had arrived, with the appearance of learned scientists free to think, speak and discuss. Instead of widening the River Fleet to solve the pollution problem, as suggested by Christopher Wren following the Great Fire of London in 1666, it was converted into the New Canal under the supervision of Robert Hooke. Further expansion and urban growth consigned the river to an underground existence. Fleet Street, named after the river, is famously known for its publishing activities, which began around 1500 with William Caxton's apprentice, Wynkyn de Worde, setting up a printing shop near Shoe Lane and with Richard Pynson setting up his publishing business near St. Dunstan's. Mostly books were printed at that time.

FIG. 5.2 Dead bodies being gathered up and loaded into carts for disposal at the time of the plague. (Courtesy of the Trustees of the British Museum)

What was Fleet Street like at the time of Tompion and Graham? They lived, socialized and pored over their wheels and cogs for almost a century between them in that busy street; and it was here that the orrery was born. Many of the original buildings are no longer there having been demolished or moved, stone by stone, to other places. But there are some that remain and some plaques and notices that record what happened at that spot. The atmosphere, however, in both senses of the word, would have been very different. During the day there would be some gaiety and buzzing of people going about their normal duties, contrasted by violence and robbery at night. Slops and waste would still be disposed of through windows at any time, and the collection of human dung in carts would still be emptied onto river barges for disposal to farmers and gardeners for manure. There were several

The Clockmaker's London

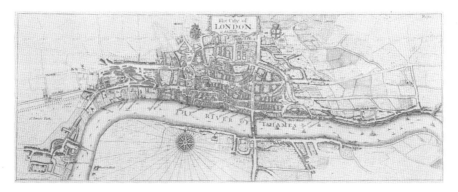

FIG. 5.3 A detail of the city of London in Queen Elizabeth I's time surrounded by fields. (Courtesy of Map in Public domain)

churches around and the priests very much regulated the behavior of society at that time.

Fleet Street rose to be one of the most important streets of London from early times when it was a main route (track) to connect the east and west of the surrounding countryside, and the street was becoming more established as the path from the commercial City of London to the political center of Westminster. It can be seen from the early print of The City of London in Queen Elizabeth's time (detail in Fig. 5.3, full map later in the chapter) that the area between, approximately, London Bridge and St James's Park was pretty much surrounded by fields. The powerful Knights Templar set the tone for the area in the twelfth century through setting up their new headquarters in the Temple Church between Fleet Street and the River Thames. After their demise, the legal profession settled in the church that evolved into the Middle and Inner Temples, two of the four Inns of Court, the other two being Lincoln's and Grey's. The church survived the Great Fire of London in 1666, although somewhat refurbished by Christopher Wren, but was heavily damaged by bombs in the Second World War.

A rough sketch of Fleet Street and its modern environs is shown in Fig. 5.4, the focus, of course, being where Tompion and Graham lived and worked. The name of Whitefriars ranks highly in the history of Fleet Street and Tompion and Graham were in the middle of it! There were four major sects of mendicant religious orders of friars originating in the early- to mid-thrirteenth

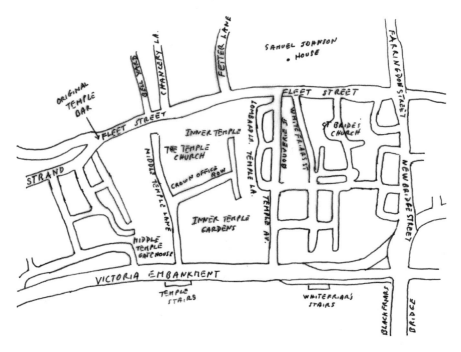

FIG. 5.4 A rough sketch of London from Fleet Street to the river. (Sketch by the author)

century, the Dominicans, founded in 1216 and known as the Blackfriars, the Franciscans, founded in 1209 and known as the Greyfriars, the Augustinians, founded in 1244 and known as the Austin Friars, and the Carmelites, founded in 1155 and became known as the Whitefriars.

The Whitefriars were from Mount Carmel in modern day Israel, but were driven from the Holy Land by the Saracens. They were allowed into England and built a small church in 1253 in Fleet Street. The area between Fleet Street and the Thames that is bounded by Bouverie Street and Temple Avenue to the west, and Whitefriars Street and Carmelite Street to the east, eventually became known as Alsatia. Following the dissolution of the Carmelites, jurisdiction of the area became uncertain but still had the status of a sanctuary. The charter granted to the inhabitants of Whitefriars in 1608 gave the area some additional elements of self-rule. Not unexpectedly, all manner of criminals, especially those fleeing from bailiffs, congregated here. The law eventually put a stop to that in 1697, which also put an end to Alsatia. Remnants

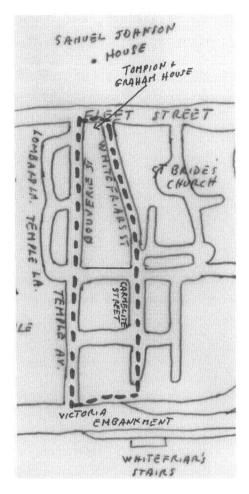

Fig. 5.5 Indication of the boundary of Alsatia. (Sketch by the author)

of the church crypt are to be found in Magpie Lane off Bouverie Street. Once again Tompion and Graham would have witnessed the historic events and violence at very close hand! Important to note is that Whitefriars Street did not have that name at the time of Tompion and Graham. It was called Water Lane. The rough sketch in Fig. 5.5, an annotated detail from the above map, shows the boundary (dotted line) of Alsatia and the location of the Tompion and Graham dwelling, right on the corner of Alsatia and right in the middle of all the trouble.

In fact, Tompion, when first recorded in 1674 to be in London (although he had probably been there since 1671) actually resided in Water Lane, further down than the corner of Fleet Street, an obscure craftsman in an obscure shop. With the clatter of dung cart wheels constantly echoing through the narrow lane, the flow of pedestrian and other wheeled traffic and, no doubt, gravity hastening the general filth down toward the river, visitors preferred to meet up with Tompion elsewhere. This would usually mean at one of the coffee houses. As Tompion became more recognized and commensurately more financially secure, he moved house and in 1675 was definitely installed at the corner of Fleet Street and Water Lane. Maybe he would stroll down to the river and cross over via Blackfriar's Bridge? Actually, no, because nothing even resembling a bridge had been constructed there at that time. Its building was begun late in the 1700s. Opened in 1769, it was originally named the William Pitt Bridge, but soon, through common usage, named to indicate the locality of the Blackfriars Monastery. How do we know so much about the layout of London and the lives of Tompion and Graham in the seventeenth and eighteenth centuries? It is a relevant diversion to recall here two particular sources; John Strype's survey and the diaries of Robert Hooke.

An early attempt to survey London was made by John Stowe (1525–1605), a merchant tailor who was admitted to its company in 1547 and who was also a writer of literary history, the 'Woorkes of Geffrey Chaucer' being his first. His meticulous Survey of London was published in 1598 and not only described the streets and buildings, but also reflected on the social conditions and customs of Elizabethan England. Further editions were published, but the most significant was that produced by John Strype (1643–1737) in 1720 and again in 1754 by others since so much had happened in the previous hundred years or so.

Several devastating bouts of the Plague, prior to the last in 1665, had occurred in 1603, 1625 and 1636 killing 10,000, 35,000 and 30,000 residents respectively. So, amazing then, that the population of London nevertheless increased from 60,000 in 1500 to 350,000 in 1660 and 575,000 in 1700. The Plague was a disaster waiting to happen because the general household waste and human detritus, with no alternative, was thrown into the streets. Disease-carrying fleas and their robust live transport, rats, thrived on the ideal conditions.

Fig. 5.6 Plaque attached to the monument near Pudding Lane, commemorating the origin of the Great Fire of London. (Courtesy of Eileen Thompson)

The Great Fire of London that started in a bakery in Pudding Lane in 1666, devastating as it was, had the result of changing London's buildings from plaster and wood to houses of stone, but not before making a little less than 90 % of its population homeless. The Monument near Pudding Lane and the attached plaque commemorating the incident are shown in Figs. 5.6 and 5.7.

People from all over the world, fleeing from famine or persecution, came to London not as wastrels but bringing with them new manufacturing and other skills; Dutch brewers, French tailors and a needle maker from Africa, who all contributed to England being one of the foremost manufacturing nations of the world. The Hugenots were a Calvinist protestant and anti-catholic group in France who were persecuted for their separation and for their sometimes violent and iconoclastic behavior toward the Catholics. The Hugenot's criticism of the Catholic Church became very severe. No wonder, then, that the Catholics replied with the St. Bartholomew's Massacre, 1572, when thousands, possibly up to 30,000, people were killed. That elicited a greater Hugenot response, even resorting to armed attacks to take over cities. The Edict of Nantes, 1592, made Catholicism the official religion of France, but gave some tolerance to the Hugenots. However, things got worse for them and in 1685, Louis XIV revoked

Fig. 5.7 The monument near Pudding Lane, full view. (Courtesy of Eileen Thompson)

the Edict of Nantes and made Protestantism illegal by the Edict of Fontainebleu. So the choice was to be forcibly converted to the catholic faith or to flee the country, the latter being the choice for more than 200,000 people. 50,000 fled to England, of which 10,000 moved on to Ireland, where they assisted Henry VIII in the dissolution (ransacking) of the monasteries. They were welcomed to their first city of settlement in England, the beautiful and historic city of Canterbury, Kent, where they carried out their weaving occupation at The Weavers. They were also recognized as some of the greatest silversmiths between 1680 and 1760, a beautiful example of their work is shown in Fig. 5.8. The rights and equality of the Hugenots in France were restored in France in 1787/89 with the Edict of Versailles and the Declaration of human rights.

As London businesses flourished, with the accompanying pressure on time and competition, there was a demand for more

FIG. 5.8 A beautiful Hugenot plate. (Courtesy of the Trustees of the British Museum)

efficient travel and transport and Hackney Carriages, later to develop into hansom cabs, thronged the streets and the wherries, river buses, packed the river. Samuel Pepys recorded that he was held up for an hour and a half while traveling in a carriage and then gave up and walked; a seventeenth-century traffic jam!!

Strype identified streets and lanes around Fleet Street in a perambulatory documentation called 'A Survey of the Cities of London And Westminster' that it is still possible to follow on the modern map. Unsurprisingly, lots of the names have changed and some disappeared but Newgate, Ludgate, Shoe Lane and Seacoal have survived, the latter being named after the coal boats that unloaded there. Nearby Newcastle Lane has a similar provenance. Fewtars Lane has become Fetters Lane but the House of Rolls is now a huge office building named Rolls House near to the Public Records Office. Fewtars Lane was a garden for the relaxation of the legal profession. The origin of Fewtars was the old English word meaning 'idle people'!

Water Lane is mentioned (now Whitefriars Street) so maybe George Graham saw Strype marching down the street with clipboard in hand. More than one hundred years earlier than Strype Stowe described Water Lane in his survey: "Then is Water lane running down by the west side of a house called the Hanging

Sword, to the Thames". The alley way connecting Water Lane to Salisbury Court was named Hanging Sword Alley after the so named house (because patrons had to hang up their sword before entering to show they meant no trouble) and even before that it was called Blood Bowl Alley after some dreadful things that happened there. Some early references to Water Lane reveal that, in the second half of the sixteenth century, walking down the Lane was not a pleasant experience. 'Water' fell on people's heads and great dung heaps sloshed down to the river. People could hardly pass through due to the piles of 'waste', and the broken and ruined pavement made it too difficult for horse and cart to access the way. What made it even worse was that it was much narrower than known today; it was not widened until after the Great Fire of London.

The small area around Whitefriars, mentioned earlier, was a hotbed of crime, bullying and sloth. A famous murder took place here. Lord Sanquhar, a Scottish nobleman, had his eye put out by accident by a fencing master, John Turner. The lord was not convinced it was an accident and, even though Turner showed genuine regret, took much time considering how to exact his revenge. He made several abortive actions to kill Turner but failed, so he recruited two thugs, Carlisle and Irving, who shot Turner in a tavern in Whitefriars. Lord Sanquhar was hanged outside Westminster Hall and the two thugs were hanged on the same day on two gibbets in Fleet Street against the great gate of the Whitefriars. There are records of many other murders, and therefore hangings, in and around Whitefriars and Fleet Street.

Many of the same streets described by Strype were previously noted by Stowe: Seacoal Lane, Windagaine Lane "and thence back again for there is no way over", Fewtars Lane, Ludgate and the House of Rolls.

In spite of the grim image of filth and smells conjured up from many records, Stowe described the provision of water for the inhabitants of Fleet Street. Spring water from the parish of Paddington had been tapped during the fifteenth century and fed through lead pipes to a point at the corner of Shoe Lane and Fleet Street to give clean water to the poor for drinking and the rich to "dresse their meats". The water was on tap by 1471. In 1478 the residents of Fleet Street were licensed to build, at their own expense, two "cesternes", one to receive the clean water and another near the prison at Fleet Bridge to receive, and dispose of, waste water. The main spring-water-receiving cesterne was enlarged in 1582. Early maps show the 'conduit' situated in Fleet Street. However, Stowe described at that time that the "cesterne, made by the men of Fleet Street for the receite of spring water at Fleet bridge", had been left to decay.

Prevalent in Fleet Street as much as anywhere in London were the coffee houses. The drinking of coffee has many legendary origins, but coffee most likely found its way from Ethiopia in the early sixteenth century to Constantinople, where it was seen as a threat to the Imams since people discussed religious matters, amongst other topics, and was banned. Taste for the bean spread rapidly throughout Europe, frequently within coffee houses, which were places to relax, gossip and debate. They were a welcome change from taverns where the usual alcoholic beverages engendered bad behavior and criminality and were hotbeds of noise and nuisance and certainly no place for women (except those for the pleasure of men).

There were, however, many taverns in Fleet Street in the seventeenth century. One of the most famous and oldest of these hostelries was the Devil's Tavern (originally called St. Dunstan's) of No. 1 Fleet Street, the site now being occupied by the bank of Messrs. Child & Co. It was even old in those days, having a recorded existence of well over one hundred years before the Tompion/Graham era. Ye Olde Cock Tavern (originally called the Cock and Bottle) at number 201 was similarly old and was frequented by Dr Johnson, Thackeray and Samuel Pepys (who frequented so many other taverns, possibly seeking entries for his diaries), although it was taken down and rebuilt over the other side of the street, number 22, to make room for a branch of the Bank of England. It is now just a replica since much of it was destroyed by fire around 1990. The Mitre, originally built on the site that is now Messrs. Hoare's Bank, number 37, was of great antiquity, even having good evidence that Shakespeare was aware of it. An even older tavern was the Castle that was to be found, at least as far back as 1432, on the west corner of Shoe Lane and Fleet Street. It was the rendezvous of the Clockmakers' Company until 1666. Ye Olde Cheshire Cheese now at 149 Fleet Street is another very old tavern that was popular in the seventeenth century. Tompion and Graham must have been utterly surrounded by taverns since there were also two more close by Water Lane, the Old Ship and the Boars Head, the latter probably dating from 1646 and stood on the site of what is now number 66. A plan exists of the Black Lion Inn in Water Lane, now Whitefriars Street. The property belonged to the Worshipful Company of Ironmongers.

Although Fleet Street was severely damaged in the Great Fire of London, it presents itself much as it was at that time. In addition to the multitude of taverns in the main street itself, there were many close by, such as the Trumpet, in lanes and streets leading into it that boasted famous people being amongst their patrons. Also the King's Head and the Pope's Head in Chancery Lane. Each of the taverns has its own illustrious or infamous history and a 'Who's Who' of fascinating celebrities of the age as their customers.

As mentioned above, the first credible records showed that coffee originated in the fifteenth century in Ethiopia, the modern country of the ancient Abyssinia and Eritrea, from where it spread north from Africa, through the Middle East and to Europe. According to Strype, coffee was first known in England in the mid-seventeenth century when Mr Daniel Edwards, a Turkish merchant, brought from Smyrna to London one Pasqua Rosee, a Ragusean youth, who prepared this drink for him every morning. This beverage style of his attracted too many inquisitive acquaintances, so he allowed his servant and one of his sons-in-law to set up the first London coffeehouse, Bowman's, in St. Michael's Alley in Cornhill, the second being the Rainbow in Fleet Street, or, maybe, the Temple Bar. The first in England, 1650, was The Angel in Oxford, which attracted many of the virtuosi from the intellectual strata, one of the first being Sir Isaac Newton. It was somewhat elitist and labeled a Penny University since the coffee was cheap and included access to a newspaper and maybe some tobacco. The exclusivity soon disappeared across the country and the coffee houses spread fast. Although there were many dissenters, not least the King for a very short time, the advantages became most acceptable. Instead of the imbibing of copious quantities of mind-dulling beers of the taverns, where no sensible business could be transacted, the coffee sharpened the senses and enabled serious

Fig. 5.9 A rare drawing (possibly created 1650–1750) of a coffee house interior that indicates the atmosphere and goings-on. (Courtesy of the Trustees of the British Museum)

debate—or to just sit in a corner and read without disruption. Unexpectedly (surely they would have preferred sensible locations for their men instead of having them arrive home roaring drunk and violent), women were against the houses and formed the Women's Petition against Coffee in 1674. The petition, at the behest of the Liberty of Venus, included "grand inconveniences accruing to their sex through the excessive use of that drying, enfeebling liquor". They were not actually banned from the establishments and Anne Rochford and Moll King (a notorious thief, pickpocket and prostitute), famous owners and women news-hawkers, delivered leaflets on their rounds. Figure 5.9 shows a rare drawing of the interior of a late seventeenth century coffee house where "You have all Manner of News there: You have a good fire, which you may sit by as long as you please: You have a Dish of Coffee; you meet your Friends for the Transaction of Business, and all for a Penny, if you do not care to spend more."

The coffee houses often became stylized such that physicians resorted to Batson's, Garraway's in Change Alley attracted commercial characters and Jonathon's and Sam's were notorious for their stock-jobbers. Jonathon's was built around 1680 on the site of the original London Stock Exchange and was the venue for a plot being hatched to assassinate William III. John Castaing posted the prices of stocks and commodities there in 1698. Women were not generally known to reside at coffee houses, but in 1848, two simply dressed and timid-looking ladies sought accommodation at the Chapter coffee house in St. Paul's Alley since it was the only accommodation they knew of from their father. The style of this venue was very much toward the publishing fraternity, and the two ladies were Charlotte and Anne Bronte, who had come to clear up some difficulties with their publishers, Smith and Elder. At their peak there were said to be as many as 3,000 coffee houses in London.

It must have been very convenient to have available so many venues to pop in to meet by arrangement or chance those with whom one would wish to socialize, have intellectual discussions or progress business. Although they were not necessarily adhered to, there were coffee house rules, printed in 1674, such as barring discussion of 'sacred things', speaking poorly of the state, no games of chance, no antagonism between customers otherwise the instigator would have to buy the offended a cup of coffee and swearing would attract a fine of 'a twelve-pence'. Indeed, great scientists, merchants, artists and playwrights could gather to discuss topics broadly and probingly without the pressure of being seen to be right or being recorded (by note-taking, not the latest 'mobile'!). There were no social class barriers ("No man of any station need give his place to a finer man") and knights of the realm and locals with common interests mingled. Hypotheses, theories and political and business contacts must have thrived in such relaxing social conditions. A major attraction was the supply of leaflets of sorts that spawned a thriving business of providing up-to-the-minute news and views, much gained from the trawling of the gossip from these venues anyway, to be read for free or at least included in the common coffee house price of one penny. As Fleet Street was the main thoroughfare between the commercial part of London and the political Westminster area, little wonder that here was the birth of the newspaper industry.

FIG. 5.10 Showing how close was Graham and Tompion's residence to the Tipperary Tavern. (Photograph by the author)

So what about the clockmakers? They certainly made plenty of use of coffee houses and Jonathon's was one of Hooke's locals since he resided at Gresham College not far away. He recorded some meetings that typified the conversations; "At Jonathon's, Sir Ch. Wren about planetary motion". On the 4 January 1680 he wrote, "At Jonathon's, Sir J. Hoskins, Tompion, Trumbol,—perfect Theory of the Heavens". Hooke seemed to spend much of his life popping in and out of, and staying at, coffee houses, as referred to in his diary, all over London. They included Man's (kept by Alexander Man) in Charing Cross, also known as the Royal Coffee house, being one of the many by royal appointment, and having a third title of Old Man's Coffee house to distinguish it from Young Man's across the street. Such a huge network of places to talk technical and arrange commissions must have been like a moth to a candle for him.

One wonders whether Tompion and Graham ever actually patronized one of the oldest Taverns in London that was right next door, The Tipperary. It was probably noisy and rough and made it difficult to concentrate on the finer details of clockmaking. However, Tompion knew exactly what he was moving into as his prior abode was just round the corner in Water Lane. Figure 5.10 shows just how close was the tavern, and its provenance is explained on the poster outside, Fig. 5.11. Because a Tompion

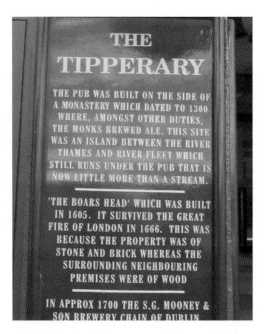

Fig. 5.11 Provenance of the Tipperary Tavern is explained on the poster outside. (Photograph by the author)

clock was included in the 1700 refurbishments it is likely that he did have a good neighborly relationship with the Tipperary or, at least, a commercial one with the new owners.

Popular as coffee and coffee houses were another taste crept into the English psyche in the mid-seventeenth century. Tea! The 'herb' was clearly known many millennia previously in China, but only filtered through to Europe much later. The drink was known to a few in England around 1660 and it was mentioned in Pepy's diaries. But it remained a rarity until the restoration, when Charles II married the Portuguese princess Catherine of Braganza. Her main attraction was her wealth, which was intended to help Charles recover from the debts left by Cromwell as well as his own flash spending habits. Catherine was a devout catholic and also didn't drink alcohol, but was very fond of tea. Having been made somewhat ill from a drink of ale she insisted on tea from then on. The ladies of the time noticed this and drinking tea became very fashionable, great as a counter to the male-dominated coffee houses. One brake on its initial popularity was the huge price since it had to be brought from India and other eastern coun-

FIG. 5.12 The Cutty Sark on public display in Greenwich, London. (Photograph by the author)

tries all the way round the Cape of Africa. The traders realized the highest price for the product brought by the fastest boats that would take many weeks for the long journey. So was born the great boats with massive sails, the Tea Clippers.

Inevitably, there was much prestige to be gained by the fastest and amongst the best of them was the Cutty Sark, the only one of the originals to have survived today. It now rests on show in Greenwich, London, on public display, Fig. 5.12. The Cutty Sark was built in Dumbarton, Scotland, on the River Clyde. She was launched on 22 November 1869 and left the Clyde on 13 January 1870, around a month before she set sail from London to China. On each of her eight voyages to Shanghai or Hankou she picked up 600,000 kg of tea. Later she would carry general cargo to Sydney, Australia, before picking up 1,000 tons of coal to be delivered to Shanghai. The opening of the Suez Canal in 1869 and the introduction of steamships gradually pushed the tall sailing ships out of the tea trade (Fig. 5.13).

Fig. 5.13 The Cutty Sark figure-head showing Nannie the witch holding bits snatched from the horse's (Meg) tail. (Photograph by the author)

Fig. 5.14 A detail of Twinings, Fleet Street, London, established 1706. (Photograph by the author)

Thomas Twining was the first to set up a tea room, in 1706, and its shop, with the same family and on the same site, still exists today at 216 The Strand, which is a few feet away from the Temple Bar where The Strand meets Fleet Street (Fig. 5.14). Surely Graham and his contemporaries must have indulged in the drink of the moment that has become the unofficial drink of England.

Before leaving Fleet Street, the home of the first orrery makers, there is one landmark to note that has been briefly referred to. At the junction of The Strand and Fleet Street, in the center of the road, stands a monument known as Temple Bar. The name arose from the Knights Templar who held much power here, and Temple Church still stands there today. The Bar developed into a pair of gates that were newly painted and hung in 1553 and were replaced

Fig. 5.15 Temple Bar etching around 1672. (Courtesy of the Trustees of the British Museum)

by more sturdy ones in 1554 on the arrival of Philip of Spain. The structure became more substantial and made of wood with a prison above and, at some time, widened to prevent the blocking of vehicles. As time moved on, the Bar began to look dilapidated and plans were afoot to replace it. Inigo Jones (1573–1652) the great architect, was appointed to the design but, although the Bar was spared from the Great Fire of London, that incident, coupled with the Great Plague, distracted focus on such relatively trivial matters. However, in 1668, soon after these disasters, minds were focused once again on the Bar, which had seen such pivotal points of history as Cromwell passing victoriously through the gates and Charles I on his way to have his head removed.

Following the usual wrangling over funds to pay for the job, Christopher Wren was given the task to design. The old one was duly demolished and the new one, made entirely of Portland stone, was completed in 1672. Figure 5.15 shows a print from an etching by an unknown person of an elevation of Temple Bar with small arches either side of a wide arch, statues in niches above and staffage in the foreground. Figure 5.16 shows a somewhat cleaner Temple Bar area, around 1770, which looks less like a sewer and thugs paradise, although there are still a couple of heads on spikes as a reminder of the times.

Fig. 5.16 A somewhat cleaner Temple Bar area, around 1770. (Courtesy of the Trustees of the British Museum)

The Bar structure remained despite many petitions for its removal because it restricted the flow of traffic. Finally, the decision was taken to remove it in 1874, stone by stone, and all 400 tons of them were placed, logically labeled and numbered, in storage in Farringdon Road. The monument was replaced by the current one, shown in Fig. 5.17, sporting the statues of Queen Victoria and the Prince of Wales with a griffin on top. After about ten years, the old stones were purchased by the brewer Sir Henry Meux, who rebuilt in Theobolds Park, Hertfordshire, in 1888 (Fig. 5.18). In 2001 the Bar was purchased and rebuilt in Paternoster Square adjacent to St. Paul's Cathedral. The return to London of this iconic structure was completed in November 2004 (Fig. 5.19).

Figure 5.20 shows a sketch of Fleet Street to summarize the best guesses of where the clockmakers lived and socialized. The letters are a selection of some houses, taverns and coffeehouses (and one item, the cesterne) where they are or are likely to have been.

FIG. 5.17 The current Temple Bar monument with the statues of Queen Victoria and the Prince of Wales and a griffin on top. (Photograph by the author)

Near Tompion and Graham's house was also the shop of George Adams (1720–1773), The Sign of Tycho Brahe's Head, probably number 60 on the south side (left on the diagram) of the street. Here, his Grand Orrery was made, now at the Whipple museum of the History of Science, Cambridge. Adams was instrument maker to the king, as was his son. A gallery of a selection of images of Fleet Street is shown in Figs. 5.21a–g.

The provenance of Benjamin Cole (1695–1766) an English instrument maker, engraver and map maker, sheds more light on

FIG. 5.18 Temple Bar rebuilt in Theobolds Park, Hertfordshire, in 1888. (Courtesy of Creative Commons Attribution-Share Alike 3.0 Unported, attribution; Mnewnham)

FIG. 5.19 Temple Bar rebuilt in Paternoster Square adjacent to St. Paul's Cathedral. (Photograph by the author)

the address 136 Fleet Street where John Rowley plied his trade under the sign of The Globe. Cole was apprenticed to Thomas Wright, and the firm of Wright and Cole operated until 1748 when Cole succeeded Wright, forming the company of Cole and

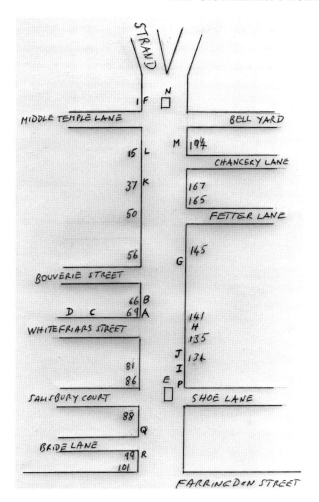

Fig. 5.20 A sketch of Fleet Street to summarize the best guesses of where the clockmakers lived and socialized (sketch by the author). *A* Tompion and Graham house, *B* Tipperary Tavern, *C* Hanging Sword tavern, *D* Early Tompion house, *E* Conduit or cesterne, *F* Devil's Tavern at no. 1, *G* The Old Cheshire Cheese, *H* Rowley house at no. 136, *I* Later Graham house between the Globe and the Duke of Marlborough's Head, *J* The Globe at no. 134, *K* The Mitre, *L* The Rainbow, *M* No. 194, The Old Bank of England tavern, previously the coffeehouse of the Cock and Bottle, *N* Temple Bar, *P* Castle tavern, *Q* The Old Bell, *R* Punch tavern

Son, which was active between 1751 and 1766. The address of his company was given as the 'Orrery adjoining the Globe Tavern in Fleet Street' that became 136 Fleet Street about 1760 and 200 Fleet Street in later years. John Troughton took over the business in 1782. Cole's business pamphlet (Fig. 5.22) reads "Mathematical

224   Orrery

Figs. 5.21 a–g Relevant buildings in modern day Fleet Street. Left to right, going down: Site of The Devil Tavern; The Old Bell; Building numbers 141 and 135 either side of the gate (Rowley number 136); Ye Olde Cock Tavern; Site of Mitre Tavern; Punch Tavern; Twinings. (Photographs by the author)

and Optical Instruments of all Sorts Accurately made according to the Best and Latest Improvements By Benjamin Cole at the Orrery next the Globe Tavern in Fleet street London."

Much of the history of London is gleaned from the study of maps. The oldest map recognizable as such is one of the whole of Britain of unknown origin but is referred to by the name 'The Gough map' since it was bequeathed to the Bodleian library about 200 years ago by Richard Gough (1735–1809), an English antiquarian. The map has been dated provisionally as 1360, although a project is underway to attempt to date it more accurately by matching the writing to known scribes. It is not a street map, but shows the names of towns with simple drawings.

There are many somewhat later maps of London, but the first with much detail appears to be an early-sixteenth century copper plate map on which subsequent maps of the time were based. It was only discovered in the mid 1990s and consisted of two panels, although many more panels were surmised to exist. A third panel was found just a few decades ago and the hunt is on for the

Fig. 5.22 Benjamin Cole's business pamphlet. (Courtesy of Public Domain. (PD-US))

remainder. A map of the City of London in the reign of Queen Elizabeth is typical of these early maps (Fig. 5.23).

A wood-cut map of London, c1550, clearly shows Water Lane beginning at Fleet Street and wending its way down to the river with Whitefriars Steps to the west and Bridewell and the emerging river Fleet to the East. Also Temple Bar as a significant pitched roof building.

John Norden's map of 1593 (Fig. 5.24) confirms the extent of London at the time. As for most contemporary maps, great importance is placed on marking the steps or boat landing/mooring places. The River Thames was the 'motorway' and lifeblood of the city. Strype's map (Fig. 5.25) of 1720 is a little clearer, and the detail, magnifying Graham and Tompion's locality, (Fig. 5.26) artistically includes the Blackfriars, Dorset and Whitefriars stairs.

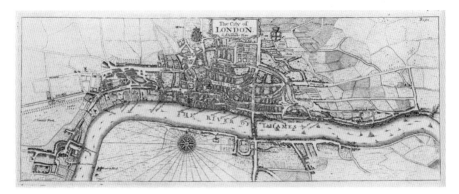

Fig. 5.23 City of London in the reign of Queen Elizabeth. (Courtesy of Public Domain)

Fig. 5.24 John Norden's map of 1593. "A guide for Cuntrey men In the famous Cittey of London". (Courtesy of Public Domain. (D-US))

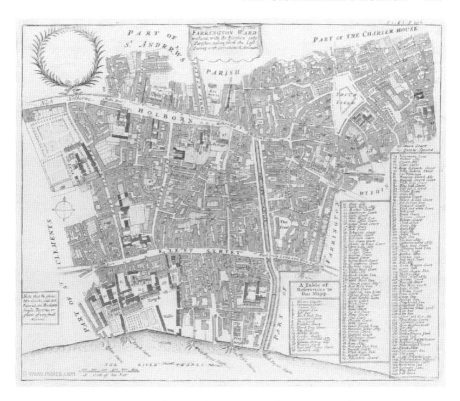

Fig. 5.25 Strype's map of 1720. (Courtesy of motco.com)

Greenwich

Greenwich, London, in particular its park and association with the River Thames, crops up time and time again during the unfolding story of not just Tompion and Graham, but clockmakers, scientists, the monarchy and other notable persons of the time. And what a year it was, 2012, for Greenwich being awarded the title of Royal Borough, hosting equestrian events of the Olympics and taking part in Queen Elizabeth's diamond jubilee celebrations. Figure 5.27, two Greenwich Views, are an etching from 1723 showing the Observatory from the south side and a panorama of London across the river with its many spires. Figure 5.28 is a view from the river, about 1770, of the Ancient Palace called Placentia, which was built by Henry V and was destroyed in the Civil Wars. Figure 5.29 shows the Royal Hospital, 1720, with the stairs by the river and fields behind.

FIG. 5.26 A detail of Strype's map of 1720. (Courtesy of motco.com)

Greenwich Park was first enclosed by King James I, who, supposedly, gave it to Queen Anne as an apology for swearing at her when she accidently shot one of his favorite dogs! It has an area of 183 acres, is the oldest enclosed royal park and slopes down from the top of a hill to the historic buildings below. Beyond, just a short distance away, is the River Thames, which can be seen in the panoramic view from the top of the hill (see gallery below). Standing in the center of the awesome buildings is the Queen's House, built in 1616 for the Queen of James, I, Anne of Denmark, although she died before completion. The old Royal Naval Hospital, amongst its other history, was opened in 1705 as a hospital for disabled and aged naval pensioners that Graham and Tompion

Fig. 5.27 Two Greenwich Views from 1723 showing the Observatory from the south side and a panorama of London across the river with its many spires. (Courtesy of the Trustees of the British Museum)

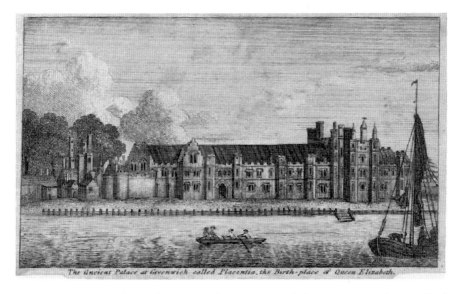

Fig. 5.28 A view from the river, about 1770, of the ancient palace called Placentia built by Henry V, which was destroyed in the civil wars. (Courtesy of the Trustees of the British Museum)

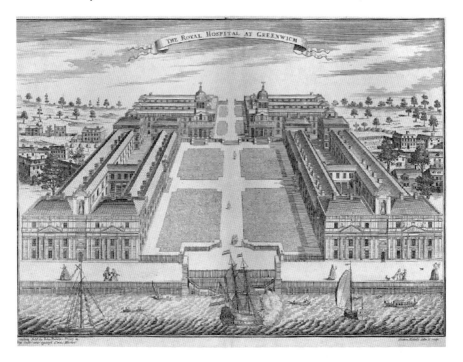

FIG. 5.29 The Royal Hospital, 1720, with the stairs by the river and fields behind. (Courtesy of the Trustees of the British Museum)

would have witnessed. There are underground tunnels, conduits, in the park that fed water down the hill to the hospital and a red brick building housing an entrance to the tunnels that still stands today. The tunnels were in existence in the seventeenth century, brick built and almost large enough to stand up in and, once again, Graham and Tompion might well have been aware of these during their regular visits. There are many plane trees, the great survivors of London smog (smoke and fog) that pervaded the city about half a century ago. They shed their bark, which reduces the build-up of toxic dust and grime (Fig. 5.30).

The observatory stands sentinel at the top of the hill, where the internationally renowned meridian dateline is marked in the courtyard. The maritime and astronomical theme is apparent around the park with two sundials, a large traditionally shaped one at the bottom of the hill in the play area and the Dolphin Sundial in the observatory rear court yard. On top of the observatory, Flamsteed House, is sited a time ball, one of the earliest public

The Clockmaker's London 231

FIG. 5.30 A London plane tree in Greenwich Park. (Photograph by the author)

time signals (1833) because clocks and watches were still too expensive for common use. In particular, it could be seen from the River Thames to assist those who sailed by. Every day at 12.55, the time ball rises half way up its mast. At 12.58 it rises all the way to the top. At exactly 13.00, the ball falls, providing a signal for anyone wishing to note the precise time. It drops at 13.00 GMT in winter and 13.00 BST in the summer. It can be seen from many vantage points around the park, from the south (away from The River) and looking west. There is also a working time ball in Deal, Kent, that was the first to be operated by a direct signal (via the electric telegraph of the S.E.Railway) from Greenwich. Sir George Airy, the seventh Astronomer Royal, supervised the installation and controlled its operation for many years. There are sixty time balls still in existence around the world, although only a few are operational. The first time ball was constructed in Portsmouth, England, in 1829. Figure 5.31 shows a gallery of images of Greenwich Park referred to above.

FIG. 5.31 A gallery of images of Greenwich Park, consisting of **a–f**. From left to right and going down: The south side of the observatory and time ball; looking west across the park toward the observatory; a panoramic view from the observatory toward London; the entrance to the old hospital tunnels; looking southward toward the hospital; view toward central London while standing on the Greenwich meridian. (Photographs by the author)

From one of the tall ships sailing down the Thames from Woolwich to beyond Greenwich, a jubilee event, a succession of historical landmarks could be seen, and photographed. A selection of images in Figs. 5.32a–d follow the route.

A special treat for the thousands of national and international spectators of the jubilee celebrations was seeing the Queen's barge on the Thames (Fig. 5.33) in a flotilla of hundreds of boats of all kinds (Fig. 5.33) sailing down the 'Royal River'. The photographs were taken toward Blackfriar's Bridge from a position a short distance away from Water Lane (now Whitefriars). Figures 5.34a–e show a gallery of images from around Greenwich Park.

A fairer hearing

Those who uncover the secrets of the world through statistics often describe a link from anyone to everyone via a chain of just six people, friends, colleagues, acquaintances and businesses. Being in the same group of like-minded enthusiasts might bring in

Figs. 5.32 a–d The plaque with facts about the river; approaching the Thames flood barrier; the domed O2 arena with the new zip wire river crossing in the foreground and Canary Wharf commercial buildings on the Isle of Dogs; the riverside view of the Greenwich buildings complex in which can be seen the tiers of grandstand seats in the background with thousands of cheering spectators from all over the world for the 2012 Olympics. The time ball can just be seen behind. (Photographs by the author)

a chance contact with a public figure, providing a short route of connection to a monarch or president. Or the route might take all six steps or more if, serendipitously, each link is remote from the target's links. And so it is that, in the telling of the history of one person, other respected (or not so respected) figures pop in and out of the story whose contribution might be small but, nevertheless, significant. A few notable or influential people encountered in 'Orrery' are worthy of a greater mention to give a wider appreciation of the backdrop to events and achievements. Some may also

234   Orrery

FIG. 5.33 Sailing down the 'Royal River'

get a fairer hearing if their previously abbreviated appearance in the text judged them harshly.

Jonathan Swift (1667–1745)

Link with Orrery: Involved in the quarrel over the Epistles of Phalaris.
   Gulliver's Travels, written by Swift, 1726 amended 1735, must be amongst the best known books of all time. He was born in Dub-

The Clockmaker's London 235

FIG. 5.34 a–e A gallery of images from around Greenwich Park. The Shepherd's Clock; tombstone plaque of Edmond Halley; the imposing statue of General Wolfe looking down from the top of the hill (that of William IV is at the bottom); a triangular sundial; a dolphin sundial

lin, Ireland, although his father, also named Jonathan, was from Herefordshire and his mother from Leicestershire in a wonderfully named village, Frisby-on-the-Wreake. Hence he is often referred to as Anglo-Irish, especially as he traveled back and forth between England and Ireland, sometimes for family reasons and sometimes for political ones. Figure 5.35 shows a bust of Jonathan Swift near his burial spot on St. Patrick's Cathedral, Dublin, Ireland.

Rather than being involved in the quarrel between Charles Boyle and Robert Bentley over the Epistles of Phalaris, he was more of an observer of a philosophical needle match between the Ancients and the Moderns. It began in France in the late seventeenth century over the question of whether contemporary learning had surpassed the knowledge of classical Greece and Rome. The 'moderns', typically Bernard le Bovier de Fontenelle, believed

Fig 5.35 Bust of Jonathan Swift near his burial spot in St. Patrick's Cathedral, Dublin. (Courtesy of Creative Commons Attribution-Share Alike 3.0 Unported, attribution; Wknight94)

that the modern age of science and reason was far superior to the superstitious and narrow world of Greek and Roman thinking. The 'ancients' argued that all that was necessary to be known was written by Aristotle, Homer, Virgil and Cicero. In the opinion of the 'moderns', modern man saw farther than the 'ancients' ever could. The tiff spread to England when William Temple, on the side of the 'ancients', responded to Fontenelle in an essay inventing two metaphors that would be used by many future authors and speakers, including Newton in his spat with Hooke. The first, which meant that modern man could only see farther because of the learning and observations of the classical philosophers, that modern man was just a dwarf standing upon the shoulders of giants! The second saw modern man as just reflected light as op-

FIG. 5.36 The diaries of Samuel Pepys. (Courtesy of Public Domain, {PD-US})

posed to the ancients who were the source of light. Richard Bentley and William Wotton, the critic, responded to the essay and the lines were drawn; Bentley and Wotton on one side and Fontenelle and, eventually, Atterbury and the 'Christ Church Wits' (Oxford) on the other. In retrospect it seems quite perverse for Atterbury to have involved Charles Boyle in the quarrel, who was an undergraduate at the time. However, Swift, from the sidelines, wrote a short but stunning satire entitled The Battle of the Books. Although written in great detail, he wisely did not give a conclusion as to who was the winner.

Samuel Pepys (1633–1703)

Link with Orrery; association with Greenwich, Flamsteed and Newton.

Pepys was born in Fleet Street, London, and was well educated in his youth. He attended Huntingdon Grammar school, St Paul's, London, and in 1650 went to Cambridge University. He bequeathed the huge collection of books he had collected, over 3,000 including his diaries (Fig. 5.36) and bookcases, to Magdalene College from where he took his Bachelor of Arts degree in 1654. Figure 5.37 shows Magdalene College in 1870; the college still exists today. Gruesomely, Pepys attended the beheading of Charles I

FIG. 5.37 The Pepys Library in Magdalene College, Cambridge, houses the diaries and other books of Samuel Pepys. (Courtesy of Creative Commons Attribution-Share Alike 2.0 Generic license; attribution: photograph ©Andrew Dunn, 10 October 2004)

in 1649. He married 14-year-old Elisabeth de St Michel, a descendant of French Huguenot immigrants, in 1655. In 1660 he became a councilor of state in Cromwell's Protectorate.

It would be easy to get a grim impression of life in the seventeenth century, with religious infighting, the Plague in 1665, and the Great Fire of London in 1666, when Pepys had to repeatedly remove his wife and his gold from London but, in his famous and meticulously detailed diaries, Pepys reveals some lighter notes such as that of the first performance in England of the seaside entertainment show Punch and Judy, which was actually in Covent Garden in 1662. Punch and Judies are now common at seaside resorts and locals fairs (Fig. 5.38). He kept those diaries for just nine and a half years, from 1 January 1660 to 31 May 1669. For his talent for administration, but no maritime experience, Pepys rose to be Chief Secretary to the Admiralty under Charles II and James II. In his role as being a most influential man in the navy, he set standards that many acknowledge made him Father of the Royal Navy. He approached the proper listing and high standards of records of armaments, stores and spares and the proper treat-

The Clockmaker's London 239

Fig. 5.38 Punch and Judy show at the local fair. (Photograph by the author)

ment of sailors with particular attention to their food. He set high standards of the understanding of mathematics and navigation and contributed much to the founding of the Royal Mathematical School at Christ's Hospital and introducing exams. In 1675 he was appointed governor.

Nevil Maskelyne (1732–1811)

Link with Orrery: associated with the 'longitude problem' and Greenwich.

Dava Sobel's book 'Longitude' blasts out from its cover "The True Story of a Lone Genius Who Solved the Greatest Scientific Problem of His Time". It is a stunning story well told, as also is the film of the same name that is based on the book. Harrison came up from humble beginnings to achieve the heights of brilliance in clockmaking. Against all odds, he proved again and again that he had created the perfect marine chronometer. His biggest enemy was the wicked Nevil Maskelyne, who tried to frustrate the poor clockmaker at every turn, almost until his death. The film presents Maskelyne glowering down at Harrison from his powerful committee of The Board of Longitude, hell-bent on sinking 'the clock' in favor of his own more practical method of Moon and star positions. That was a great stance to provide a great film and also introduced so many to the history of longitude and navigation.

But was that unfair to Maskelyne? The bicentenary of his death was celebrated not long ago, which spawned some revisiting on the protracted episode of longitude. His character does not seem to be at one with descriptions of vindictive attacks on anyone. He was a great character, convivial and liked by all. Jean-Baptiste-Joseph Delambre, a notable French astronomer, referred to Maskelyne as 'le dieu de l'astronomie' and others, astronomers and philosophers, found in him a brother. He had a wide circle of friends and acquaintances abroad and became a member of the French Institute. He never suddenly invented anything, but worked hard at diffusing astronomical knowledge and advice. He was sent to St Helena, a volcanic island in the south Atlantic, to observe the transit of Venus in 1761,, but the event was clouded out and Maskelyne spent much time making accurate relative measurements of the Moon's position in the sky. In fact, he was responsible for the development of the Nautical Almanac, which included also the ephemeris of stars, invaluable to mariners at sea, and still used even until recent times. Continued delving into his biography consolidates the opinion of his meticulous and mathematical pursuit of astronomy and his likeable personality. Perhaps his story might not have made such a good film, but was it justifiable to characterize Maskelyne as such a villain? Maybe not! Was he a genius at mathematics? Probably!

Sir Cloudesley Shovell (1650–1707)

Link to Orrery: The longitude prize.

It is easy to hate Sir Cloudesley Shovell from the reading of several anecdotes of the longitude story. He commanded the boat,

Association, and the fleet returning from Gibraltar in 1707 with an iron fist. Not only did he ignore the warning of proximity to the rocks of the Scilly Isles from an ordinary seaman, and that the longitude of the ship's position was not as Shovell thought, he immediately had the seaman hanged by the yardarm for inciting mutiny, so the book and film would have it. Not unreasonably, since it was the incident rather than the person that was relevant to the longitude history, little more about Shovell is contained in any of the tellings and re-tellings of the Scilly Isles catastrophe. Although this incident was merely recounted to starkly illustrate the lack of navigational knowledge at the time, which led to the longitude prize, nonetheless the image set in the reader's mind of Shovell's brutality and arrogance leading to the death of 2000 souls taints his reputation.

His unforgettable first name was the surname of his maternal grandmother Lucy Cloudisley. His father, John Shovel, was descended from a family who had property in Norwich, but were not wealthy. Note the alternative spellings for his name. Use of the surname for a forename is not rare. Indeed, Charles Boyle's son by Margaret Swordfeger was Boyle, but all four illegitimate children adopted the surname of Swordfeger—otherwise he would have been called Boyle Boyle! In 1691 Cloudesley married Elizabeth, Lady Narborough, the widow of his former commander. Through her, he had two stepsons, who at the ages of 23 and 22 also died in the sinking of the HMS Association on the rocks of Scilly. Shovell's wife, the former Lady Narborough, is buried in St Paulinus Church, Crayford (Fig. 5.39) where the memorial (Fig. 5.40) also mentions her husband.

Even by the age of just 23 Cloudesley had seen much action in battles at sea and rose through the ranks. He had also studied navigation. In serving under Sir John Narborough, a distant relative, then transferred to The Harwich, he took part in and led successful actions against the Barbary Pirates, for which he received awards of money and a gold medal from King Charles II. In a letter from the Admiralty, Samuel Pepys recorded the King's satisfaction with Shovell's actions. He continued to rise in rank and eventually spent nine years as captain of his own ships fighting the Barbary Pirates. He greatly distinguished himself at the Battle of La Houghe, under Admiral Russell, by being the first to break

FIG. 5.39 Sir Cloudesly Shovell's wife, the former Lady Narborough, is buried in St Paulinus Church, Crayford. (Photograph by the author)

through the enemy's lines aboard HMS Prince, although severely wounded here. He returned to sea and action in battles and promotion kept on coming; he was made first Admiral, then Knight of the Realm. He was promoted to Admiral of the Blue in 1696 and Admiral of the White in 1702 by Queen Anne, of whom he was a particular favorite. During the Wars of Spanish Succession, in which he fought under Admiral Rooke in 1704, he cooperated to capture the town of Gibraltar. The island was ceded to Britain in 1713. He continued to fight in many successful battles and in 1706 was appointed Commander-in-Chief of the British Fleets whilst at Lisbon. At one time he also served under Prince Eugene. Being faced with the heroic life-long deeds to defend his country, how could anyone believe that Sir Cloudesley Shovell was anything other than a brilliant and brave patriot.

FIG. 5.40 The memorial to Sir Cloudesly Shovell's wife, in St Paulinus Church, Crayford. which also mentions her husband. (Photograph by the author)

Located a few minutes walk from Charing Cross station and close to the Embankment is to be found one of London's most unusual pubs, The Ship & Shovell (Figs. 5.41 and 5.42), which is on two sides of the street with the cellar below linking the two sides.

Sir Richard Steele (1672–1729)

Link to 'Orrery': naming of the orrery.

In the context of the orrery, Steele is noted as having mistakenly in a lecture credited Charles Boyle, fourth Earl of Orrery, with the design and invention of the geared model of the solar system. He went so far as to credit himself with the naming of the model as an orrery, which may be true. All around him wrote to

FIG. 5.41 The Ship and Shovell pub showing premises on both sides of the street. It claims to be the only such one in London. (Photograph by the author)

FIG. 5.42 Showing the two inn signs of the divided Ship and Shovell pub. (Photograph by the author)

correct his mistakes, but was he such a bumbling fool, or did he make just one of those errors that most would admit to at some time in their life? Such a passing cameo in a story gives little in the way of clues.

He was born in Dublin, Ireland, a member of the Protestant gentry. He was educated at Charterhouse School in Godalming, England, and spent short times at Christ Church and Merton Col-

FIG. 5.43 The name of Sir John Vanbrugh crops up many times during the seventeenth and eighteenth century as an architect or dramatist. He lived in the castle shown here situated on the outskirts of, and overlooking, Greenwich Park. (Photograph by the author)

lege, Oxford, before joining the Household Cavalry to support King William's wars against France. He rose through the ranks to achieve captain, but eventually disliked the army and left in 1705. He was soon appointed to a position in the household of Prince George of Denmark, consort of Anne of Great Britain. He was briefly a member of parliament but was more in favor when the protestant George I came to the throne and knighted Steele. Steele was a co-founder, with his friend Joseph Addison, of the magazine The Spectator.

Although he was a prolific and successful writer, he was also heavily involved in politics, a dangerous preoccupation in those times. He was a member of the famous Whig Kit-Kat club that possibly received its name from the keeper of a pie house, Christopher Catling, hence Kit-Kat or Kit-cat, and met at various venues in London. The objectives were initially said to be to unwind after a hard day's work, but the objectives were soon crystallized to support a strong parliament, a limited monarchy, resistance to France and the protestant succession to the throne. Some of the characters in the squabble over the Epistles of Phalaris were members, as was Sir John Vanbrugh (1664–1726), English architect and dramatist and designer of Bleinheim Palace, whose castle (Fig. 5.43) still remains on the fringes of Greenwich Park and the Royal Observatory with a Blue Plaque on the wall (Fig. 5.44). The

FIG. 5.44 The life of Sir John Vanbrugh is commemorated in a Blue Plaque on the castle wall. (Photograph by the author)

fine buildings that accommodated the Greenwich hospital, designed by Wren, Vanbrugh and Hawksmoor, can be seen at Greenwich today. Toward the end of his life Steele retired, in 1724, to his wife's homeland of Carmarthen in Wales and died and was buried there within St Peter's church in 1729.

Sir Jonas Moore (1617–1679)

Link to orrery: Lent money to Tompion; principle driving force behind the establishment of the Royal Observatory, Greenwich.

Moore has been cited frequently throughout this work, but he was such a catalyst for all things good in science and astronomy, it is worth emphasizing a couple of instances relevant to the Graham/Tompion era.

Two Tompion regulators were made for the Greenwich Observatory in 1676 (when Graham was just three years old) and were paid for out of the personal pocket of Moore.

Hooke had much respect for Moore, or his money, the respect being confirmed in Hooke's diaries through the use of Moore's full name and title; no Tomkins or other demeaning label as for others: "Much Discourse with him [Tompion] about watches. Told him the way of making an engine for finishing wheels, and a way how to make a dividing plate; about the torme of an arch; about another way of Teeth work; about pocket watches and many other things'. Having 'Sent for quadrant from Tomkins' and a note to '.. send him 25 sh' Hooke, on th 15th May, presented the quadrant to Sir Jonas Moore".

He used his wealth and influence, accumulated from the huge project of draining the Great Level of the Fens, to become a patron of astronomy and the main driving force behind the establishment of the Royal Observatory of Greenwich (ROG). Moore was disappointed with the productivity of Flamsteed's work at Greenwich and threatened in 1678 to stop Flamsteed's salary if the productivity didn't improve. In fact, Flamsteed was so meticulous that he had still not published his data after forty years!

John Flamsteed (1646–1719)

Link to orrery: First Astronomer Royal at Greenwich.

Reading through the history of astronomy of the late seventeenth and early-eighteenth centuries, the perceptions of the character and ability of Flamsteed can be confusing. Was he a goody or a baddy? Was he helpful to others, and who were his friends? The reality lies in the squabbles that were common at that time. The age of the enlightenment had opened the door to free thinking and freedom of invention, accompanied by the pursuit of credit, recognition and financial rewards. A passing mention has already been made of the Newton/Hooke contention and Flamsteed's apparent withholding of star data, but a little knowledge of Flamsteed's life encourages some sympathy for his struggles.

A great deal of information became available through the publication of Flamsteed's own copious notes that he made throughout his life, but only through reading other relevant biographies in addition can a historical balance be made between impressions arising from the statements of eminent scientists. His 'diaries' were first published more than 100 years after Flamsteed's demise by Francis Baily, Vice-President of the Royal Astronomical Society in 1835.

His notes begin, wonderfully, "I was born at Denby (five miles from Derby), in Derbyshire, August 19, 1646, at a quarter of an hour past seven at night". As for many people of the time, he suffered greatly from maladies and describes how, at the age of fourteen he was "visited with a fit of sickness, that was followed with a consumption, and other distempers". Nevertheless, he determined that it would not interfere with his learning and was, therefore, very much self-taught and became recognized as an astronomer. In 1670 he traveled to London to become acquainted with eminent scientists and Sir Jonas Moore in particular, who became his friend and patron. Through Moore he was appointed to the position of King's Astronomer and paid a paltry £100 from the Board of Ordnance, paltry because promises of the supply of instruments never materialized, so all were made and repaired by himself or given by Moore. The initial enthusiasm by royalty had waned. He moved into Greenwich observatory in 1676 and attempted observations with the few instruments available, but he needed a mural arc, and the one he built was inadequate. One Abraham Sharp, said to be best mechanic and calculator of his time, constructed the required instrument, and it is from 1689 that Flamsteed's reliable observations began.

He may have been the first person to see Uranus, although he didn't identify it as anything more than a new star. His main contribution to astronomy was a new and more accurate method of determining the coordinates of stars to update the compilation of Tycho Brahe that had been produced more than a century earlier. His work he entitled Historia Coelestis; it was eventually published after his death by his wife, with help, under the title Historia Britannica Coelestis. Figure 5.45 shows a portrait of Flamsteed as a frontispiece to Historia Coelestis, 1712.

Flamsteed's notes describe his initial and friendly communications with Newton, who promised not to reveal data to anyone else, just use them to develop his hypotheses. It appears that Newton may have passed on some information to Halley, and Flamsteed challenged Newton, during which letters he impugned Halley, a good friend of Newton. Newton appears to have continued to bombard Flamsteed for more and more star data, whose illness was again referred to in his notes; "This request of Mr Newton for more observations, caused an intercourse of letters between us, wherein I imparted to him about 100 more of the moon's places; which was more than he could reasonably expect from one in my circumstances of constant business and ill health. The year following (1695), I was ill all the year with a periodical headache;

Fig. 5.45 A flamboyant portrait of Flamsteed as a frontispiece to Historia Coelestis, 1712. (Courtesy of the Trustees of the British Museum)

which was carried off in September by a violent fit of my dreadful distemper, the stone. In the mean time, frequent letters passed between me and Mr Newton, who ceased not to importune me (though he was informed of my illness), for more observations; and with that earnestness that looked as if he thought he had a right to command them; and had about fifty more imparted to him. But I did not think myself obliged to employ my pains to serve a person that was so inconsiderate as to presume he had a right to that which was only a courtesy". Flamsteed's dislike of Halley hastened the conflict between the two, which was domi-

FIG. 5.46 A bust of John Flamsteed at the Museum of the Royal Observatory, Greenwich. (Courtesy of Klaus-Dieter Keller, Germany. Public Domain)

nated by Newton, who had much more influence in eminent circles and at court.

The demeaning of Flamsteed had an unfortunate effect on a hard-working amateur astronomer and physicist, Stephen Gray (1666–1736), who had a habit, through necessity, of working for no pay, including as assistant to Desaguliers. As a friend and assistant to Flamsteed he was targeted by Newton, thereby diminishing any prospects of recognition of his genius. He also seems to have been dogged by ill health throughout his life.

Whatever the squabbles and accusations, Flamsteed is often referred to as 'the Father of practical astronomy in England'. He may well have withheld data, but he may well have had good reason to do so. Figure 5.46 shows a bust of John Flamsteed at the Museum of the Royal Observatory, Greenwich.

There are many more players that could be labeled as the supporting cast of the Orrery—another time! Even many of those mentioned have much more in the way of discoveries or significant contributions to history to be recognized. For example, Edmond Halley was the first to publicly demonstrate in 1693 predictable longevity tables with which to calculate the premium someone should pay to purchase a life-annuity.

6. Modern and Orrery Times Compared

If You Were There

Unpredicted as well as expected natural events have always been facts of life, as has social and technological evolution. Sitting in our comfy workshops and laboratories it is easy not to have in our minds the conditions under which precision microengineers toiled around the time of Graham. Electricity and gas had not been exploited for the benefit of easy living and the dim (compared to what we have now) light from candles or oil burners must have caused some eye strain at least. The fumes from the burning oil, wax and fat might have hung heavy in the atmosphere, aggravated by the low ceilings often prevalent at the time. Even the common wood-burning open fires could have contributed to the fug. The streets of London became just a little brighter around 1680, when an oil lamp was hung outside every tenth house and lit for part of the year.

But let's imagine another unfortunate burden of the times. Scientists and politicians may deliver facts and hypotheses and pontificate about the effects by humans on current global temperatures but, obviously, the Sun is, and has always been, responsible for most of what happens here on Earth. How could it be otherwise? The Sun is known to go through a sequence of cycles of activity of approximately 11 years, or 22, to take full account of the magnetic reversals after each 11-year cycle. At the peaks of activity, called maxima, the Sun's surface erupts more often than at other times, often ejecting huge plumes of plasma many times as massive as the Earth. These are known as coronal mass ejections, CMEs, which, when they crash into the Earth's magnetic field, generate huge waves of direct current (DC) electricity along power cables built to host alternating current (AC) blowing circuit breakers, overheating and melting the windings of trans-

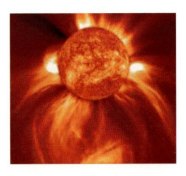

Fig. 6.1 The powerful Sun squirts out masses of energy. (Courtesy of NASA)

formers and causing massive failures of electrical distribution systems. One great magnetic storm occurred in 1989 in Quebec, when the entire electrical power system was put out of action. Coronal mass ejections and the radiation associated with masses of fast-moving solar and cosmic particles can disrupt computer equipment, which is bad news for the ever-increasing use of such systems on the ground, in the air and in space. Figure 6.1 captured the moment when around a billion tons of matter was ejected from the surface in a CME.

An indicator of solar activity easy to observe (only under the strictest of safety conditions to avoid being blinded) is the appearance of areas of magnetic turmoil on the surface, known as sunspots, which look dark relative to the bright surround. The spots can be huge, often many thousands of kilometers across, and can herald warmer weather for us. Conversely, lack of the appearance of sunspots can indicate colder weather. In recent times, during October 2003, there was a particularly large sunspot event. Figure 6.2 shows the activity recorded by telescope; many of the spots covered an area greater than the diameter of the Earth. A second image (Fig. 6.3) captured at increased magnification shows the extent of the violent eruptions. Although Chinese records of solar observation go back at least 2,000 years, it was the invention of the telescope around 1610 that kick-started the creation of a record of sunspot numbers in western scientific culture.

These records show that there was a period, 1645–1715, about the time of Graham and Tompion, when there was a particular dearth of spots. This is now referred to as the Maunder Minimum.

Modern and Orrery Times Compared 255

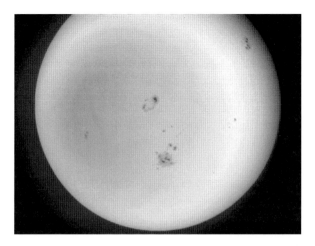

FIG. 6.2 A rash of sunspots in October 2003. (Photograph by the author)

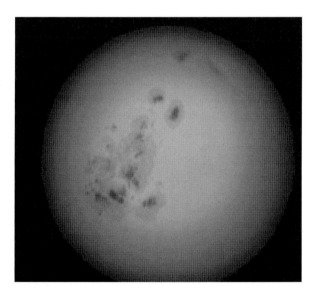

FIG. 6.3 October 2003 sunspots captured at increased magnification showing the extent of the violent eruptions whose size would swallow the Earth. (Photograph by the author)

As one might expect, this coincided with a very cool period (aided by increased dust from global volcanic activity such as from Papua New Guinea in 1660), which is referred to as the Little Ice Age. It was so cool, in fact, that Europe and North America suffered unusually cold events such as the freezing of rivers not

FIG. 6.4 The frozen Thames in 1677. (Painting by Abraham Hondius (1631–1691))

used to being frozen most notable being the freezing over of the River Thames. So thick and of long duration was the ice, that people were skating on the solid river and setting stalls to ply their businesses (Fig. 6.4). So the poor clockmakers of the time also had to endure the freezing weather. It is of interest that although the cycle of sunspots now seems to be routine and dependable, though somewhat variable, there are indications from solar seismology observations that the precursors to the formation of spots might not be forming. The conclusion to that might be the lack of future spots, heralding another 'Maunder Minimum' and mini-ice-age from about 2020—or it might not. The prediction of 'space weather' is constantly improving, but not 100 % there yet.

And yet, in spite of having to cope with the cold and possibly dim lighting, instruments turned out by so many clock- and scientific instrument makers were not hurried, primitive objects with just enough work to produce functional equipment for the intended use. They were ornate, often overly so, to announce to the world the expertise and dedication expended on their unique products—as all were. No mass production at that time for complex works of art. The attention to detail of the science was also heroic, especially to advance the quest for ever more accuracy

Modern and Orrery Times Compared 257

Fig. 6.5 A beautiful aurora seen from the International Space Station. (Courtesy of NASA)

and precision in the scientific scales (micrometers) and the times (clocks). No wonder instruments from the time of Tompion, Graham, Hooke, Mudge, Ferguson and many more fetch huge prices at auctions, since they are as much adornments for the showcase mantelpieces and aristocratic polished reception room tables as examples from history.

In the year 1716 all in London looked up to see huge lights blanketing the sky; massive and brilliant lights. Everyone came out into the streets of London, which became crowded. So bright was the vision that one witness reported that due to the brightness of the light, as for a full Moon, she saw people walking across Lincoln's Inn fields, and their shadows, but there was no Moon that night. Many imagined that they heard gunshots and smelled powder. Political events were blamed. The Jacobites proclaimed it to be an omen against the punishment of their leaders. It was, of course, the spectacle of an aurora, a storm of charged particles hitting the Earth that usually cause the lights to appear toward the poles (although never actually AT the poles). Figure 6.5 shows a stunning band of green aurora taken from the International Space Station over Nebraska in January 2012. Since aurorae are usually linked to an eruption of sunspots, as a feature of the Sun's activity, there would have been a dearth of aurora during the Maunder Minimum and they would therefore not be witnessed in the lifetime of many watchers. A second reason for concern was that the scientists of the day could not explain the phenomenon. The effects of the Earth's magnetic field were little understood at that time, and they could not have known about the Sun's violent particles, which, as it happens, would not have affected them as much as now since computers, electricity grids and satellites are vulnerable. George Graham was certainly into magnetism, even creating a sideline in the business of making magnetic rods for compasses. The top astronomers were encouraged to explain the lights, and Edmund Halley was called upon by the Royal Society to do so.

Halley made a grand effort putting forward his previous theory for wandering terrestrial magnetism in which he suggested that the Earth possessed four magnetic poles, one pair situated at the ends of the axis of an outer magnetic shell and the other at the extremities of the axis of an inner magnetic core. He postulated that the space between the core and shell was filled with a luminous effluvium, a favorite theoretical device of the time, which escaped at the behest of the motion of a changing terrestrial magnetic field. Scientists were floundering to explain the phenomenon and one must not be too judgmental when the extrapolation was made from Halley's hypothesis that the Earth was hollow. Indeed, theories expanded on it to suggest that there were many layers to the Earth, as for an onion! Of course, then came the natural lead into the cavities being populated by aliens, or at least, people. Amazingly, this perception lasted, in some quarters, right up to the early twentieth century, when Hitler pronounced that he had a massive army holed up in the spaces just ready to come out and do his worst. Naturally, the common wisdom throughout those centuries was that the underground entrance was at a location in the Arctic, but some identified an entrance in caverns near the Mississippi river. Graham and his contemporaries would have been well into hearing all about these natural or supernatural explanations and maybe deliberating their own hypotheses in the coffee houses. Although aurorae are not usually seen as far south as London, they do occur at this latitude.

Once more on the theme of the weather, the great storm of 1703 was devastating. It was the worst natural disaster ever experienced. The barometric pressure may have sunk to as low as 950 millibars, although only 973 was actually recorded. Trees were blown down (4,000 oak in the New Forest, England), floods were caused that drowned people and farm animals and one boat was found 15 miles inland. Buildings were damaged and chimney pots fell and killed many passers-by. The lead roof of Westminster Abbey was blown away and the Eddystone Lighthouse destroyed. Ships were blown all over the place such as in the Thames, where 700 were heaped together in the Pool of London. Thirteen royal ships were lost along with an estimated 10,000 sailors. HMS Association, later to be wrecked off the Scilly Isles and captained by the Admiral of the Fleet, Sir Cloudsley Shovell, was blown from Harwich to Gothenburg in Sweden. Reports recorded the damaging emotional and psychological effects of the storm. Tompion and Graham must have witnessed and feared all this!

Relationships were not always as friendly as it might seem from reading the history. Certainly, Honest George made life and collaboration much easier for many, and each guild such as the Clockmaker and Blacksmith companies stuck together (mostly), but there was much squabbling. Popular history would have it that Hooke was always accusing everyone of purloining his ideas, with some justification; Newton hated Hooke, Hooke hated Newton; many got at Flamsteed for keeping his results to himself, again, with some justification. Several Astronomers Royal variously had a go at each other and promptly changed sides, and Maskelyne got at everyone who dared to disbelieve his belief that mapping of the skies was the only route to knowledge of longitude at sea, and just dare to fall out with those who had the monarch's ear and it was straight to the tower!

Careful following of the fashionable religion was definitely recommended, again, to avoid the risk of estate confiscation or off to the tower or off with the head! But that was difficult. Five monarchs spanned the lives of Tompion and Graham and each had their own religious raison d'être.

Tompion was already ten years old when Charles I was beheaded, which heralded the next ten years of continued uncertainty of government and religion. There was some parliamentary tolerance of minority religions such as Presbyterians, Baptists and Jewish synagogues, but Catholics were excluded from such toleration. The Levellers wanted a governance system fair to all and the Diggers wished for it to be even more so. At the outset of the

Interregnum, the Puritans imposed an austere lifestyle, banning theater and gambling, but allowing the more 'virtuous art' of opera. Cromwell was brutal to the Catholics in Ireland killing many and even selling some as slaves to Jamaica. Henry Cromwell, the fourth son of Oliver Cromwell, adopted a more conciliatory attitude to Catholics such as lifting the ban on their living in towns. With the help of Roger Boyle, first Earl of Orrery, one of Cromwell's major victories in Ireland was diplomatic rather than military, persuading the Protestant Royalist troops in Cork to change sides and fight with the Parliament. Richard Cromwell, who took over following his father's death in 1658, relinquished his position as Protector without much hesitation and opened the door to restoration of the monarchy. Toward the end of the Interregnum, Parliamentary generals Charles Coote (first Earl of Mountrath, 1610–1661) and Richard Boyle seized the strong points in Ireland in preparation for the return of Charles II. It all sounded like a good time to keep your head down and play with cogs and wheels!

Under the rule of Charles II it was still a good idea to be Church of England. Under the Clarendon Code, named after the first Earl of Clarendon, English historian and statesman and grandfather of Mary II and Queen Anne, nonconformists (not conforming to the Church of England) had a hard time. They were excluded from public meetings and from teaching—and more. The code also insisted that all municipal officials must take Anglican communion and use the recommended prayer book. Charles II was, essentially, a Catholic and attempted to bring in religious tolerance as a shoehorn to Catholicism. This led to the Exclusion Crisis, exclusion being the aim of Parliament to exclude any Catholic heirs to the throne, in particular because the heir presumptive, James, Duke of York, was a Roman Catholic. The Tories were opposed to this exclusion while the 'Country Party', soon to be named the Whigs, supported it. Charles sided with the Tories, and when a plot (possibly fabricated) to kill Charles and his brother was discovered, killings of Whigs followed. Charles dismissed Parliament and ruled alone until his death in 1685.

In spite of efforts to the contrary, a Catholic, James II, came to the throne, having, according to the Earl of Lauderdale, all the weaknesses of his father without his strengths. He married Anne Hyde before he died, who had converted to Catholicism, although both had kept this secret until forced to reveal it when faced with having to swear an anti-Catholic oath. Samuel Pepys commented: "the Duke of York in all things but his codpiece is led by the nose by his wife". His reign was a very busy one, but was mostly characterized by struggles between Protestants and Catholics. A group of nobles, The Immortal Seven, invited William of Orange, James' son-in-law, to England. He accepted this invitation with 12,000 men, landing at Torbay in Devon in 1688. James fled and the crown was offered jointly to James' daughter, Mary, and her husband, who then became William III. Both were protestant.

Was this the end of the Catholic challenge and religious squabbles? Oh no! After many years of William and Mary entering a veritable contract with the people of the country, a constitutional monarchy evolved with John Locke, the philosopher, asserting that kings rule not by divine right, but by popular consent, as embodied in a 'social contract'. By this contract individuals give up certain 'natural rights' for 'civil rights' and if a government fails to protect these civil rights it may be justly overthrown. Very well in theory but up pops James again, this time in Scotland, and these Jacobites, as they were known, had some success in battle although they were crushed soon after. Still nibbling at the heels of 'Dutch Billy', as William was nicknamed, James himself landed in Ireland supported by France's Louis XIV. The Irish were looking forward to victory and getting their own back after the Cromwell experience. The Jacobites took control of almost the whole island. William fought back and, after William's victories at Boyne and Aughrim, James fled back to France.

The major religious upheavals of the previous reign seemed to be absent from that of Queen Anne, sister to Mary and cousin to William, and just maybe citizens could relax a little. Even Scotland was on side sufficiently to create the state of Great Britain. Anne was the last of the Stuarts because the Act of Settlement barred any but protestants to the throne. That excluded over 50 Catholic blood relatives leaving the Hanoverian George I to reign from 1714–1727. Jacobites tried to overthrow George but all attempts failed. Most of the controversies of George's reign were over money. Interest on government bonds and their value spiraled out of control, which led to the Bubble Act, which imposed rules on the trading of bonds. The act led to the value of bonds

Fig. 6.6 Portrait of Robert Walpole (1676–1745) in 1740. (Workshop of Jean-Baptiste van Loo (1684–1745)

spiraling out of control again, but this time downward, and huge losses were made by bond holders. No evidence could be found that any of the major orrery clockmakers of the time suffered 'Bubble losses', unlike Eustace Budgell, Charles Boyle's friend, who drowned himself. Circumstances allowed the brilliant Robert Walpole (Fig. 6.6) to be appointed First Lord of the Treasury and Chancellor of the Exchequer. He was, effectively, the first Prime Minister. Since the Tory's leanings toward Jacobitism had become so unpopular the Whigs were in power in parliament for many years and Walpole became ever more powerful. He virtually ran the country and George concentrated on international affairs—wars—and running 'his other country', Hanover. Unless they were involved in financial losses, the clockmakers and scientists might well have been able to concentrate on their work

FIG. 6.7 Purchase of Christian captives from the Barbary States

during this first Hanoverian reign. George certainly supported the Enlightenment and allowed his critics to publish without severe censorship.

Like his father, George II (who reigned from 1727–1760) was increasingly less involved in the matters of running the country, which he left to Parliament. Walpole directed domestic policy and, after the resignation of his brother-in-law Townshend in 1730, also controlled George's foreign policy. The leaders of countries usually had to prove themselves in battle and were always going to war, George II being the last monarch to actually fight in a war and, incidentally, the last king of England to be born abroad. On the continent armies were continually swapping allegiances to fight former allies. Many scientists and courtiers in London and other cities were accused of being collaborators for foreign monarchs. Much of the future problems was the constant raising of huge amounts of money to support war efforts.

Possibly worst of all for some was the fear of enslavement. Pirates and governments from North Africa, the Barbary Coast, sent 'missions' to collect slaves of any race and any religion (Fig. 6.7) to keep topping up required population targets to maintain the viability of towns and pander to private owners. One estimate of slaves taken from Europe between 1530 and 1780 was 1,250,000, many of them from the South of England and Ireland. British sailors and British and Irish citizens were terrified of capture. No wonder such efforts and expense were invested in sorting out the

264　Orrery

Fig. 6.8 Printing press from 1811. (Courtesy of Creative Commons Attribution-Share Alike 3.0 Unported, attribution; photographed in the Deutsches Museum, Munich, Germany (GFDL-self))

Barbary Pirates by such brave and able mariners as Sir Cloudesley Shovell and others.

A more promising feature of the time was the innovation that drove so much invention of social and technical development in the sixteenth and seventeenth centuries, printing!

Books were produced from centuries before but they were onerous to prepare individually by hand. The first major step to an early form of a printing production line was based on existing screw presses. It was in part of the Holy Roman Empire, Germany, that Johannes Gutenberg (1398–1468) a blacksmith, goldsmith, printer and publisher, invented the screw press, which perfected the printing process through all its stages by adapting existing technologies. He devised the hand mould and the rapid creation of moveable type, Fig. 6.8 showing an engraving of the process from 1568. This first mechanization of bookmaking led to the first mass

FIG. 6.9 1568 printing process with "puller" and "beater". (Courtesy of Public Domain; author Jost Amman (1539–1591))

production of books, a single press being able to produce 3,600 pages per day. A printing press from 1811 is shown in Fig. 6.9. There were now best-selling books issued in their hundreds of thousands. All this dissemination of information underpinned the explosion of scientific, artistic and cultural expansion in the Age of Enlightenment, also called the Age of Reason, from around 1600 to 1750. In 1620 Francis Bacon, the English statesman and philosopher, wrote that typographical printing has "changed the whole face and state of things throughout the world".

In music, two types of keyboard were known around the time of Tompion, the clavier, based on the striking of strings, and the harpsichord (Fig 6.10), producing sound through the plucking of strings using quills. Maybe the early watch and clockmakers had access to such instruments to while away their leisure time, if they had any! Of course there were many other musical

266 Orrery

FIG. 6.10 This beautiful harpsichord is the work of two celebrated makers: originally constructed by Andreas Ruckers in Antwerp (1646), it was later remodeled and expanded by Pascal Taskin in Paris (1780). (Courtesy of Creative Commons Attribution-Share Alike 3.0 Unported, attribution; Gérard Janot)

instruments. The Italian Antonio Stradivari (1644–1737) produced what are generally considered to be the greatest violins ever; they command huge sums of money on the rare occasions they come up for sale. A modern scientific investigation conducted to ascertain why the violins were so marvelous indicated that the wood used for their production was from trees that grew during the cold period of the Maunder Minimum, when slower growth led to denser wood.

Did Tompion and Graham attend concerts given by the revered composers and pianists of the time, Scarlatti (1679–1750), Vivaldi (1678–1741), Handel (1685–1759) and Telemann (1681–1767)? Maybe some of the later clockmakers listened to Richard

Mudge (1718–1773), distinguished for his compositions for, and performances on, the harpsichord. Richard was the brother of Thomas Mudge (1715–1794), who was apprenticed to George Graham. So that gives the answer. Music was around and accessible to the clockmakers of the age.

Much has been written about the achievements of great men, scientists like Newton and Hooke and clock- and orrery makers like Tompion and Graham, but what about the ladies? Were there any women at the practical bench of cogs, wheels and orreries? A search for women in the clockmakers guild records a surprisingly large number! From 1714 to 1725 six were apprentices bound to the clockmaker George Tyler and his wife, Lucy; Mary Darby, Rebeckah Fisher, Catherine Jackson, Hannah Campleshon, Elizabeth Newton and Eleanor Mosely, who earned her own freedom and had apprenticed to her seven other ladies from 1727 to 1739. Further delving reveals that many women wished to start their own business but within London this was not allowed unless they were 'free of the city'. They therefore became bound (apprenticed) to an often unrelatedly named guild such as the clockmakers, milliners or haberdashers.

The makers of clocks and orreries, apprentices and members of the many guilds were clearly intelligent but were they educated, and if so, how? After all, this was the Age of Enlightenment! The Romans installed education but, as one quote has it, "when the Romans left, so did civilization and education". St Augustine, when he arrived in England in 597, made a good start with creating grammar schools to teach Latin to would-be priests and schools to teach song ready for participation in cathedral choirs. It is not surprising, therefore, that the first grammar school was probably founded in Canterbury, on Augustine's route into the country. The principle of education was to train for entry into the clergy, medicine or law and would include grammar, logics, astronomy, arithmetic, geometry and music. However, this comprehensive range was frowned upon by the higher authorities as enabling a scholar to read heretical and subversive material. Therefore much of the teaching was actually limited to Latin and literature in preparation for entry into the church. There was much progress and good intentions over the next few centuries in bringing education to lesser mortals than just the likes of the clergy and politicians and many schools were founded. It has to be admitted that all the subjects were focused on the church, for example, mathematics and astronomy were constantly discussed in relation to the precise date of Easter.

It was mentioned in an earlier chapter that a system of payments existed in the late seventeenth century to circumvent the common activity of robbery, another area of progress at the time orreries developed. Goldsmiths took on the new service of money lending and transfer and Tompion used the services of Sir Richard Hoare at the sign of the Golden Bottle at the end of Fleet Street near Temple Bar. Tompion gave cash to Hoare; Hoare gave Tompion a piece of paper; Tompion then gave or sent his creditor the piece of paper that was exchanged for cash with Hoare.

This system had replaced the original one of trusting the Royal Mint when Charles I seized the gold in 1640. Although the gold was repaid by Charles II, trust in the Royal Mint had gone and the bankers were still goldsmiths. Once more, Samuel Pepy's records give some examples. In 1667 Alderman Edward Blackwell changed Dutch money for him and "discoursed with him about remitting of this £6,000 to Tangier, which he promised to do by the first post." The goldsmiths retained their previous business in dealing with plate; as Pepys recorded "called at Alderman Blackwell's and there changed Mr Falconer's state cup, that he did give us this day, for a tankard, which came to £6. 10s. 0d at 5s. 7d. an ounce, and 3s. 0d. in money, and with great content thence away to my brothers."

In 1690 the Battle of Beachy Head, when the English were thrashed by the French, was the last straw, and led William III to resolve to rebuild the navy to a strength sufficient to restore dominance at sea. As money was lacking a project was set up in 1694 to invite investment at the great rate of return of 8 % to raise £1.2 million for the King. In order to induce subscription to the loan, the subscribers were to be incorporated by the name of the Governor and Company of the Bank of England. The bank was given exclusive possession of the government's balances and was the only limited-liability corporation allowed to issue bank-notes. The money was raised in 12 days when people flooded to invest, the money going into building a navy. This gave jobs for all; sailmakers, carpenters, farmers for the provision of food, blacksmiths

Tomlins, Joseph	London	Son of Samuel Tomlins, late Citizen and Sadler (of London)	-	27	860	700	154
Tomlins, Thomas		Son of Samuel Tomlins, late of London, Sadler	John Gibson	23	724	200	134
Tompion, Thomas	Fleet Street	Clockmaker	-	29	957	1,000	169
Tonson, Jacob	London	Stationer	-	6	142	500	27
Torriano, Nathaniel	London	Merchant	-	20	621	500	115
Torriano, Susanne	London	Widow	Alexander Terriano	20	620	2,000	115
Tothall, William	St. Martins in the Fields	Apothecary	-	36 42	1189 1381	200 300	205
Tourton, Nicholas	London	Merchant	-	39	1292	2,000	222
Towne, Leonard	London	Haberdasher	-	20 46	618 1505	200 100	114
Townsend, Senior, John	Winchester Street, London	Soapmaker	-	13	410	500	76

FIG. 6.11 Thomas Tompion, an original investor, £ 1,000, of the Company of the Bank of England. (Courtesy of The Bank of England museum)

and so on. A great navy was formed and so were the foundations for the British banking system.

A list of all those original investors is maintained at the Museum of the Bank of England in London. And guess who's on there (Fig. 6.11) amongst the 1267 others? Thomas Tompion, a Clockmaker of Fleet Street good for £1,000! Clearly Tompion's business was doing well to be able to lend such an amount and his financial acumen was also intact. Unsurprisingly, George Graham is not on the list as he had a deep-seated loathing of any kind of bank and was much too young anyway.

The list gives some insight as to the trades of the region; stationer, apothecary, haberdasher, soapmaker, skinner, clothworker, mathematician, minister, Lord Mayor of London (who gave £3,000), maltster, haberdasher, several spinsters, merchant (who gave £4,000), chirurgeon (a surgeon), cordwainer (a shoemaker of fine leather shoes), collarmaker, whitster (bleacher or whitener), girdler, shipwright, goldsmith, fringemaker, several knights and.... Their Majesties King William and Queen Mary, (£ 10,000).

One might guess at the origins of other names such as Stephen du Thoit, a weaver of Canterbury, quite likely a scion of the Hugenots since they were weavers residing in Canterbury in the early days of their immigration (Fig. 6.12).

Three clockmakers also appear on the index to the book of the subscriptions:

John Ebsworth, clockmaker of London; £1,000. John Ebsworth was apprenticed in February 1657 to Richard Ames and

Fig. 6.12 Huguenot weavers' houses at Canterbury. (Courtesy of SuzanneKn.)

made a Freeman in April 1665. He was Master of the Clockmakers' Company in 1697 and died in 1699 only a few years after he had contributed the money to the King.

Fromanteel, clockmaker of London, £200. The immediate reaction might be to think of Ahasuerus Fromanteel, a renowned clockmaker of the seventeenth century but, just a minute, he died in 1693 and the subscriptions to the King were in 1694! A more careful look at the list reveals that the Fromanteel actually named was Abraham. Ahasuerus Fromanteel married Maria de Bruijne in 1631 and together they had eight children, four of whom became clockmakers themselves, one of which was Abraham, who lived in London. Further inspection of the Fromanteel entry shows that he donated two lots, not just the £200, but another £300 (Fig. 6.13).

Edward Stanton, clockmaker, London, £100.

Edward Stanton (1642–1715, approximately) was Free of the Clockmakers' Company in January 1662/63 and soon after established his own business taking on fourteen apprentices between 1664 and 1705. He was made Assistant to the Clockmakers' Company in 1682, Warden in 1693 and its Master in 1697. In 1699 he oversaw John Ebsworth's will. Stanton regularly attended the Clockmakers' Company until 1715, when he most likely died.

Modern and Orrery Times Compared 271

Frets, John	Middle Temple	Esquire	-	37 9 24	1211 248 780	100 200 100	47
French, William	London	Girdler	-	25	816	200	146
Fromanteel, Abraham	London	Clockmaker	-	13 38	406 1260	300 300	95
Frostin, James	Whitehall	Esquire	-	23	743	2,000	134
Fuller, Millicent	Wife of Thomas Fuller, D.D., of Hatfield, Herts.		Edward Mundy do.	26 30	823 989	1,000 400	147

FIG. 6.13 Inspection of the Fromanteel Bank investment entry shows that he donated two lots, not just the £ 200, but another £ 300. (Courtesy of The Bank of England museum)

Shipley, Samuel	London	Fishmonger	-	41	1339	100	229
Shipman, John	London	Merchant	-	38	1233	4,000	212
Shirley, Judith	Preston, Sussex	Dame	James Bateman	46	1520	75	255
Short, Daniel	Hackney, Middlesex	Merchant	-	14	428	1,000	79
Shovell, Sir Clowsly	Goodman's Fields	Knight	John Hill	21	660	1,000	120
Showers, Sir Bartholomew			Gibbons Bagnall	41	1361	500	232
Simon, Susan	London	Widow	James Augustus Blondel	34	1128	100	195
Simon, William	Lyons Inn	Gentleman	-	4	98	2,000	20
Simpson, Humphry	London	Merchant	- William Methwen	4 20	99 628	500 1,500	20
Singleton, John	London	Silkman	-	29	931	500	165

FIG. 6.14 Sir Clowsly Shovell, an original investor, £ 1,000, of the Company of the Bank of England. (Courtesy of The Bank of England museum)

Also named on the list is Sir Clowsly Shovell, Goodman's Fields, Knight, £1,000 (Fig. 6.14). In addition to his role in history, there is often a discussion as to exactly how his name was spelt. It would be a reasonable guess that the entry within the catalogue should be the correct one, i.e. Clowsly.

Extracts of the index of investors above are amazing enough, but to sit in the Bank of England archives in front of the original ledger of over 300 years old brings history to life. Figure 6.15 shows the first page of the ledger stating the purpose and dated 21 June 1694. Some entries can be hard to read as they are quite faint and required enhancement. Figure 6.16 shows enhancement of Tompion's entry and Fig. 6.17 an enlargement of part of the original; 'Tho Tompion of Fleet Street, London, Clockmaker do subscribe one thousand pounds'. The Bank's use of Britannia as the corporate seal led 100 years later to the nickname of The Little Old Lady of Threadneedle Street' by the cartoonist James Gillray.

Around this time, 1696, Sir Isaac Newton became warden of the Mint, responsible for investigating counterfeiting. He became

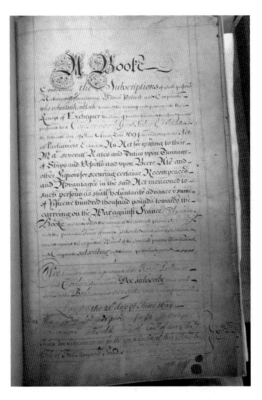

Fig. 6.15 The first page of the Bank of England ledger stating the purpose and dated 21 June 1694. (Courtesy of The Bank of England museum)

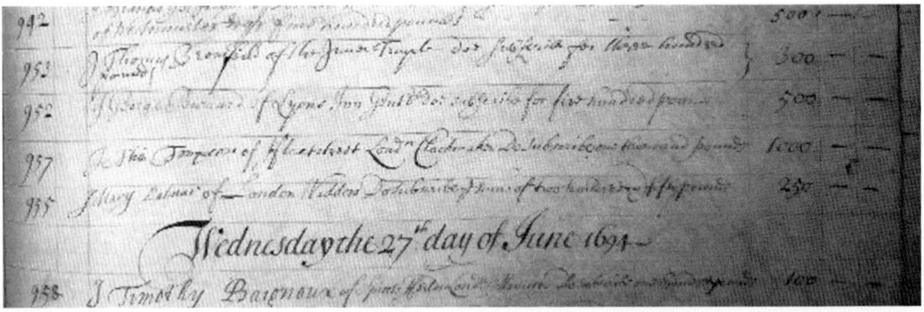

Fig. 6.16 Enhancement of Tompion's entry in the Bank of England ledger. (Courtesy of The Bank of England museum)

Master of the Mint in 1699 and remained so until his death in 1727. Figure 13.18 shows the Royal Mint coining room with screw presses, tools and Mint workers in 1808 (Fig. 6.18).

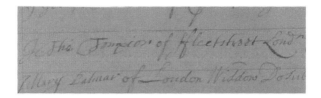

FIG. 6.17 An enlargement of part of the original; 'Tho Tompion' entry. (Courtesy of The Bank of England museum)

FIG. 6.18 View of the Royal Mint coining room, with screw presses, tools and Mint workers in 1808. (Courtesy of the Trustees of the British Museum)

It would have helped William at the time to have accurate maps with which to plan war. Some of the greatest maps were created by Pieter Mortier (1661–1711) an eighteenth century mapmaker and engraver from the Northern Netherlands. Figures 6.19a, b and c, show a particular chart that appeared in one part of Mortier's 'Neptune François', titled 'Cartes Marines a l'Usage des Armées du Roy de la Grande Bretagne'. The nine charts of this section, all engraved by Romeyn de Hooghe, one of the foremost artist/etchers of the period, was described by Koeman as the "most spectacular type of maritime cartography ever produced in 17th century Amsterdam". Mortier's motives in the production of this atlas were to flatter the Dutch king, on the British throne since the Glorious Revolution of 1688, William III,

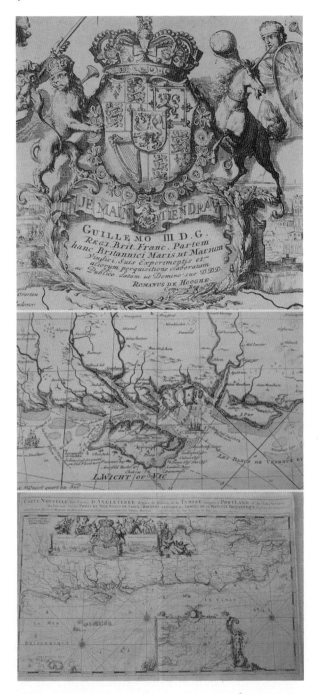

Fig. 6.19 a–c Three photographs from one map from Mortier's "Neptune François", titled "Cartes Marines a l'Usage des Armées du Roy de la Grande Bretagne" engraved by Romeyn de Hooghe, dated 1693. (Photographs by the author with the permission of Liz Lucy)

to whom it is dedicated. The unprecedented size of the atlas and the use of artists such as de Hooghe were not cheap and it has been called the most expensive sea atlas of the period, "intended more as a show-piece than something to be used by the pilots at sea". These particular specimens of the map are special in that they have not been artificially colored, as some others have, in the hope of increasing their appearance—and value.

So many ideas and suggestions for fundamental theories flew out of the minds of scientists from this age of enlightenment that it is not easy to cover them all to hint at the life and times of the beginnings of the orrery. The origin and formation of the Earth was a great target. For instance, the Danish bishop Nicolas Steno (1638–1686) was a pioneer in anatomy and geology and hypothesized that sedimentary strata had been deposited in former seas and fossils were organic in origin. Edmund Halley suggested in 1701 that the salinity and evaporation of the Mediterranean Sea could be used to determine the age of the Earth. In spite of the free thinking of the time, scientists were always looking over their shoulder to make sure they didn't upset the religious powers-that-be. With that in mind, William Whiston published in 1696 the most widely accepted theory (then) of the Earth's beginnings in his tome 'A New Theory of the Earth' when he used Christian doctrine to 'prove' that the great flood had actually occurred and that the flood had formed the rock strata of the Earth. There were some references to the supposition that water arrived here on comets.

The first comet to have been discovered through a telescope was the Great Comet, or Kirch's Comet, of 1680 that passed extraordinarily close to the Earth, just 0.42 AU. The comet is famous for being used by Newton to test and verify Kepler's laws. It is immortalized in a painting by Lieve Verschuier (1627–1686; Fig. 6.20). Some of the observers are seen to be using a Jacob's Staff, an instrument for determining angles especially at sea to find the ship's latitude.

And let's not forget the chemists! Carbon, sulphur, iron, copper, silver, tin, antimony, gold, mercury and lead were all known since antiquity. Phosphorus, 1669, cobalt and platinum, 1735, zinc, 1746, and nickel, 1751, were all discovered during the lifetimes of Graham and Tompion.

Fig. 6.20 The Great Comet of 1680 over Rotterdam. (Painting by Lieve Verschuier)

What Was It All About?

George Graham, Thomas Tompion, John Rowley, Charles Boyle, Prince Eugene; they all played their vital role in the thread of a story leading to mankind being presented with a superbly artistic and precise visual aid to the understanding of our solar system, astronomy and science. The story has spanned so many ages, discoveries, countries, battles, political and royal squabbles, renowned scientists and literary figures. It is fortunate, even exhilarating, that there was a happy ending to the pursuit of the two proto-orreries. They could so easily have been lost or destroyed. The trail highlighted the best and the worst of the characters of those involved; the generosity of Graham, the lack of reverence by Prince Eugene's niece for the collected treasures, treachery at court, the patronage of science by Charles Boyle and the brilliance and dedication of so many involved. The surrounding history illustrates how the participants lived and the adverse conditions in which they had to work. Many even had to worry for their lives and key players were often popping back and forth to the Tower as they had to guess which leader, religion or monarch to support. Get it wrong and off to the Tower again!

Modern and Orrery Times Compared 277

Fig. 6.21 The etching/engraving of the Tower of London from across the river, 1715. (Courtesy of the Trustees of the British Museum)

The image (Fig. 6.21) is great to view as an illustration of the Tower—and then pass on—but, as for many images in this work, there is more, much more to be gleaned from it. The detail created by the engraver is astounding and missing it would be to miss so much of the feeling for the time of 1715, the age of George Graham and the dawn of the orrery. Firstly, the banner to the left (Fig. 6.22). The engraver, Sutton Nicholls, lived in London (1668–1729) and was also a printseller, draughtsman and globemaker (there seemed to be a lot of those at the time) and a member of the weavers company. His father lived in Eltham, Kent, possibly a Hugenot who commuted to The Weavers in nearby Canterbury? The banner reveals that there was a tavern in Aldersgate called the Half Moon. A quick internet search shows that he lived in, or worked from, a 'Cake-shop by ye Halfe Moone' in 1703. He was also "against the George Inn and the Angel" also in Aldersgate. He was at many other places in and around the area with names reminiscent of the clockmakers; Three Compasses, Two Globes, The Gilded Ball and The Crown and Sugarloaf. So there are a few

Fig. 6.22 Banner magnification of Tower of London from across the river, Fig. 6.21. (Courtesy of the Trustees of the British Museum)

Fig. 6.23 Scroll magnification of Tower of London from across the river, Fig. 14.1. (Courtesy of the Trustees of the British Museum)

more bits of assistance for a researcher into the location of London taverns and businesses. The scroll banner (Fig. 6.23) is a key to the numbered parts of the engraving, '12' identifies the Traitor's Tower, that we might have been able to guess due to having

Modern and Orrery Times Compared 279

FIG. 6.24 Magnification of Tower of London image, Fig. 6.21, from across the river, to show the wherries, taxis of the river. (Courtesy of the Trustees of the British Museum)

FIG. 6.25 Magnification of Tower of London image, Fig. 6.21, from across the river to show passers-by going about their daily routines. (Courtesy of the Trustees of the British Museum)

the only barred windows, which was previously called St Thomas' Tower and later Traitor's gate.

In the expansion of an area of the river (Fig. 6.24) can be seen the design of the many 'wherries' transporting their properly-dressed fares and on the quayside (Fig. 6.25) passers-by going about their daily routines, kids playing against the fence, man on a horse leading a dog and two horses pulling a single-axel cart. There is so much more to be revealed by expanding every part of the image as there is for Fig. 6.26, produced a few years later (1737) with the different types of wherries (Fig. 6.27) and the larger boats that appear to be scaled-down models of the ocean-going galleons (Fig. 6.28).

And, three centuries later, has the scene changed? A photograph taken in December 2012 (Fig. 6.29) from approximately the same vantage point reveals that the Tower of London stands as solid as ever. The towers, castellations and walkways have endured, all would be recognizable by a time-traveler from those historic days of Tompion, Graham and Boyle. Even the entry to Traitors

FIG. 6.26 A later view, 1737, of the Tower of London from across the river. (Courtesy of the Trustees of the British Museum)

FIG. 6.27 Magnification of Tower of London image, Fig. 6.21, from across the river with the different types of wherries. (Courtesy of the Trustees of the British Museum)

Gate would be frighteningly familiar (Fig. 6.30)! What would have been a heck of a surprise would be the vision of the riverside from where the photograph was taken (Fig. 6.31). The building material, shape and size would be quite a shock, but which will last longest, the rock-solid Tower or the piles of weird windows?

With our modern social and technical advances it is sobering to think that such precision engineering necessary for the production of accurate instruments was carried out to the backdrop of the stench of streets, disease and the dangers of lawlessness. On the brighter side, it must have been great to be able to hop from one coffee house to another in the knowledge that someone somewhere would be looking forward to discussing your particular topic or obsession or just chilling out over the latest issue of local news and gossip. No deafening pop music to

Fig. 6.28 Magnification of Tower of London image, Fig. 6.21, from across the river with scaled-down models of the ocean-going galleons. (Courtesy of the Trustees of the British Museum)

give it extra 'ambiance' or 'cinema experience'! Even the ladies had their first 'newspaper', The Ladies Mercury, that was, in fact, just a single sheet. It was published on 27 February, 1693, but lasted only four weeks.

It is apparent that as soon as a particular invention had occurred or advancement had been made, the news and action spread rapidly for others to push forward with the technology. Hardly had the proto-orrery been in the news than clockmakers and scientists all over the country and the rest of the world were developing and improving the instrument at a great rate. Many used it to present their talent for drumming up trade. Their shop bay windows must have been wondrous to browse.

The painting shown in Fig. 6.32 by Joseph Wright of Derby (1734–1797) was produced around 1766. It was a work of art entitled 'A Philosopher Giving a Lecture on the Orrery in Which a Lamp is put in Place of the Sun'. It is an example of the technique of chiaroscuro that refers to strongly contrasting light and dark that is intended to bring out the volume or depth of the work. But to many that's all it was, a piece of art. To some it was an abomination of science. But why? There were several reasons. The people in the picture appear to be posing unrealistically and not being instructed,

FIG. 6.29 Three centuries later, a photograph taken in December 2012 from approximately the same vantage point reveals that the Tower of London has changed little. (Photograph by the author)

FIG. 6.30 A modern view of the entry to the Tower of London through Traitors Gate looks frighteningly familiar. (Photograph by the author)

at least no one seems to be paying attention. The natural philosopher and mathematician, John Rowning (1701–1771), a Fellow of Magdalene College, Cambridge, had stern words to say in 1743:

"Artificers generally erect upon the Ecliptic some Semicircles to represent some of the principle circles of the Heavens. But this is wrong and tends to Confusion because these circles are only imaginary and arising from the 'apparent' Motions of the

Fig. 6.31 A modern vision of the riverside from where the photograph, Fig. 14.9, was taken. (Courtesy of Eileen Thompson)

Fig. 6.32 A somewhat controversial painting at the time entitled 'A Philosopher Giving a Lecture on the Orrery' by Joseph Wright of Derby around 1766

heavenly Bodies, ought to have no place in the Orrery". Benjamin Martin's complaint (in 1771) was that these hybrids were a waste of money and "Stand in Need of none of the useless, expensive and cumbersome Embellishments of art".

Some even argue against the orrery itself, denying it to be a useful demonstration of Newton's theory of gravity since the orbits were perfectly and unrealistically circular. However, it was probably a great idea of Graham's as something novel, which reflected his great interest in science and astronomy, increasingly, it seems, at the expense of his clock- and watch-making. It demonstrated Graham's expertise in precision engineering and appealing presentation of objects. Indeed, the models immediately appeared to be items for the cabinets of collector's. The potential for education of the masses (at least beginning to reach lecturers, teachers and gentlemen in addition to royalty and nobility) about the Solar System soon caught on. The fact that orbits were circular was unlikely to detract from a lesson on movement around the Sun that was definitely contrary to the Ptolemaic principle that had held so dogmatically for so long. It is just possible, of course, had there been another young Hooke or similar, movements and gears with cams could have been constructed to rise to the challenge of perfect orbits, as was designed for clocks and the equation of time. Nevertheless, other challenges would have presented themselves such as the relative angles of the orbits and the unrealistic possibility of correct relative sizes and distances of the bodies. As the great Rupert Gould wrote in describing his refurbishment of the Rowley orrery:

"It will be noticed that Graham only designed this machine to show the motions of the Sun, Earth and Moon. By doing so he gained an advantage in point of scale, A little reflection will show that it is, in any event, a practical impossibility to make a model Solar System to exact scale. In the present machine, for example, the sun is a globe about 3 in. in diameter. On the same scale, the Earth ought to be about 26 ft. away, and about 1–27th in. diameter. The Moon would be roughly 7/8 in. from the Earth, and only 1–100th in. diameter." And, as Jon Darius, (1948–1993, astronomer and one time curator of astronomy at the British Science Museum) said, "for the Earth to be in the right relation to Rowley's 3 in. Sun it would have to be the size of a grain of sand a bus-length away".

The name 'planetarium' has been used confusingly and elicited the following passage:

> Planetaria enjoyed popularity, from the seventeenth century onward, as devices to demonstrate Sun-centered cosmology. Many models only show the Earth, Sun, and Moon, indicating that the

phenomena associated with these bodies—day and night, the seasons, eclipses, and lunar phases—were considered of particular import. In general, those models that included the planets were intended to provide only approximate representations of astronomical motions. The relative sizes of, and distances between, planets were not accurately represented, nor were the post-Keplerian elliptical orbits of the planets; the scale of most planetaria requires that the orbits resemble circles. Further, even in the age of Newton, the planetary machines show mean motions and fixed orbits, without attempting to illustrate perturbations or deviations due to the gravitational effects of bodies other than the Sun. Planetaria were not intended to teach mathematical theories of astronomy. Generally speaking, a planetarium demonstrates the relative motions and positions of the astronomical bodies and is primarily designed for teaching. From the first half of the 20th century, the term 'planetarium' has often described optical systems used with a projection instrument inside a domed theatre. However, the term is also routinely used to describe mechanical, non-optical demonstrations of astronomical motions.

If only the Graham and Tompion orrery could talk! What splendor it had seen or been associated with in its continental travels. In 1722 Charles VI, Holy Roman Emperor authorized the construction of a permanent home for the library in the Hofburg palace, after the plans of Leopold I. The wing was begun by Johann Berhard Fischer von Erlach. The most valuable items to go into the library were those in the collection of Prince Eugene with its 15,000 volumes of valuable books from France and Italy to add to the 200,000 books that had already existed. There was, however, criticism that the library only served to represent rather than encourage the search for knowledge, so the collection was supplemented with numerous scientific works by Doctor Gerard van Swieten, physician to Maria Theresia, and his son Gottfried, who introduced a system of card-indexing. This facilitated the updating of the inventory.

The Prunksaal, State Hall, is the central structure of the old imperial library and part of the Hofburg palace, outside of which is a statue of Eugene on horseback (Fig. 6.33). Located in the magnificent hall of the Prunksaal (Fig. 6.34) are marble statues of emperors with the statue of Emperor Charles VI in its center.

Some questions and observations pop up in the telling of the story of the orrery. It was often quoted that orreries and other

FIG. 6.33 A statue of Eugene on horseback outside the Hofburg palace, Austria. (Courtesy of Giovanni Dall'Orto, January 2004)

creations were 'sent' to Rowley. But he only lived across the road! Perhaps the roads were too mucky or dangerous even for that small distance without a commercial delivery man. There seems to be uncertainty for some as to whether the Graham model or the Graham and Tompion model was intended for Prince Eugene. It was the second, the Graham and Tompion. We may never know if Tompion was aware of Graham making the first proto-orrery or whether he actually encouraged him or tolerated the work. It does seem that Tompion was sometimes impatient with regard to Graham's output of clocks and watches. Did Tompion and Graham work together on the Graham and Tompion proto-orrery, or did it just bear Tompion's name because that was the house rule? Any mechanism that left the property had to have Tompion's name credited.

Mention of orreries in the eighteenth century crops up all over the place. The minutes of the Spalding Gentlemen's Society (SGS), 1761, refer to an orrery made by Thomas Hawkes that is still in the society's possession. Hawkes also described in detail his observations of the transit of Venus from Norwich. Coffee houses have much to answer for, or to their credit. The founding of the SGS, one of the oldest learned societies, was spawned through informal meetings in a coffee house in the Abbey Yard, Spalding, Lincolnshire, to discuss antiquities and to read The Tatler, the newly published London periodical.

FIG. 6.34 Nationalbibliothek Wien; Prunksaal, center of the old imperial library. (Courtesy of Pe-sa.)

One question needs to be answered yet, and it is like the 'elephant in the room'. It concerns the Antikythera mechanism. It was so sophisticated, complex and a demonstration of an awareness of the motions of the heavens that it is astounding that nothing else like it existed all those two thousand or so years ago! Indeed, nothing like it was seen for many hundreds of years later. There should be examples all over the place. Where are all the others or intermediates leading to it? It is most unlikely to be a hoax since models have been created to mimic its build and operation. It clearly is the ultimate orrery.

Appendix 1: A Select Timeline

1292	Richard of Wallingford, English astronomer/horologist born
1336	Richard of Wallingford, English astronomer/horologist died
1386	Salisbury cathedral clock built
1410	Prague clock installed
1473	Nicolaus Copernicus born
1524	Roger Boyle born in Herefordshire but lived in Kent, England
1525	John Stowe, surveyor of London, born
1532	Nicholas Oursian, French clockmaker first recorded
1543	Nicolaus Copernicus died
1546	Tycho Brahe born
1564	Galileo Galilei born
1566	Richard Boyle was born in Canterbury
1576	Roger Boyle died in Preston, Faversham
1590	Nicholas Oursian died
1600	Lantern clocks became common
1601	Tycho Brahe died
1602	Galileo Galilei discovered pendulum swing independent of amplitude.
1605	John Stowe, surveyor of London, died
1610	Galileo Galilei describes moons associated with Jupiter
1629	Christaan Huygens born
1633	Samuel Pepys born
1635	Robert Hooke born
1639	Thomas Tompion born
1642	Sir Isaac Newton born
1642	Galileo Galilei died
1643	Richard Boyle died
1644	Ole Christensen Rømer born
1646	John Flamsteed born
1650	Sir Cloudesley Shovell born
1650	Duke of Marlborough (John Churchill) born
1656	Professor Edmund Halley born
1657	Christaan Huygens patented pendulum clock
1657	Robert Hooke invented the anchor escapement
1660–1669	Samuel Pepys created his private diaries
1662	Dr Richard Bentley, Master of Trinity College, Cambridge, born
1663	Prince Eugene Francis of Savoy-Carignan-Soissons born
1666	Great Fire of London
1667	Jonathan Swift born
1668	John Rowley born
1672	Sir Richard Steele born
1673	George Graham born
1674	Charles Boyle born

Appendix 1: A Select Timeline

1675	John Flamsteed appointed first astronomer royal at the founding of the Royal Greenwich Observatory
1677	Stephen Hales born
1693	James Bradley, the third Astronomer Royal, born
1693	John Harrison born
1694	The beginnings of the Bank of England
1695	Christaan Huygens died
1695	Cylinder escapement designed by Thomas Tompion and patented by himself, Edward Barlow and William Houghton
1695	Beginning of the battle of Epistles of Phalaris, Charles Boyle
1698	Pierre Louis Maupertuis born
1701	Anders Celsius born
1701–1714	War of the Spanish Accession
1703	Robert Hooke died
1703	Samuel Pepys died
1704	Duke of Marlborough at Battle of Blenheim
1707	Sir Cloudesley Shovell died on the shores of the Scilly Isles
1709	Duke of Marlborough and Prince Eugene at Battle of Malplaquet
1710	Ole Christensen Rømer died
1710	New St Paul's Cathedral, designed by sir Christopher Wren, completed
1710	Proto-orreries produced around this time
1712	John Rowley makes the first tellurium to be called an orrery after his patron Charles Boyle, 4th Earl of Orrery.
1713	Thomas Tompion died
1715	Thomas Mudge born
	George Graham's dead-beat clock escapement
1719	John Flamsteed died
1721	Larcum Kendall born
1722	Duke of Marlborough (John Churchill) died
1722	George Graham made significant improvements on Cylinder escapement
1727	Sir Isaac Newton died
1728	John Rowley died
1729	Sir Richard Steele died
1730	John Harrison created design and drawings for first marine chronometer, H1
1731	Charles Boyle died
1732	Eustace Budgell, 'Memoirs of the Life and Character of the late Earl of Orrery'
	Reverend Nevil Maskelyne born
1736	Prince Eugene died. His huge fortune, buildings and collections passed to Princess Anne Victoria of Savoy, the daughter of his oldest brother who promptly sold the lot for knock-down prices without record keeping
1736	John Arnold born
1741	Anders Celsius simultaneously detected a magnetic storm
1742	Dr Richard Bentley died Professor Edmund Halley died
1744	Anders Celsius died
1745	Jonathan Swift died
1747	Abraham-Louis Breguet born
1749	Thomas Earnshaw born
1751	George Graham died
1753	Johann Georg Nesstfelt completed the repair of English 'orrery'
1754	Thomas Mudge invented the detached lever escapement

Appendix 1: A Select Timeline

1759   Pierre Louis Maupertuis died
1761   Stephen Hales died
1762   James Bradley, the third Astronomer Royal, died
1763   William Jones, clockmaker, born
1767   John Pond born
1775   Detached escapement created by John Arnold
1776   John Harrison died
1780   Detached escapement modified by Thomas Earnshaw
1794   Thomas Mudge died
1795   Larcum Kendall died
1797   James Ferguson born
1799   John Arnold died
1811   Reverend Nevil Maskelyne died
1823   Abraham-Louis Breguet died
1829   Thomas Earnshaw died
1831   William Jones died
1836   John Pond died
1851   Sir George Airy redefined the exact meridian position
1867   James Ferguson died
1884   Greenwich meridian agreed to be the official prime meridian
1890   Sir Harold Spencer Jones born
1930   G&T proto-orrery purchased by a dealer, Jesse Myer Botibol
1937   Rupert Gould describes the restoration and mechanism of Rowley orrery
1948   G&T proto-orrery sold at Sotherby's auction sales in, Lot 174
1960   Sir Harold Spencer Jones died
1974   Derek J de Solla Price published "Gears from the Greeks: the Antikythera mechanism"

Appendix 2: Glossary

Armillary	An armillary has other names such as spherical astrolabe and armil and consisting of a spherical framework, usually metal, of rings centered on the Earth. They represent objects in the sky (celestial sphere) and/or lines of celestial longitude, latitude and the ecliptic. It differs from a celestial globe that is a smooth ball produced to map the constellations.
Astrarium clock	A clock that displays more than the usual time. It includes astronomical information such as mean time, sidereal time, the phases of the moon, motions of the Sun and known planets and can display tide tables. The first to be described was one by de Dondi in 1364.
Astrolabe	An astrolabe is, essentially, a calculator and has been known since ancient times. Its mechanism is based on an inclinometer and has been used for astrologers, navigators and astronomers to determine time and latitude and predict the positions of the Sun Moon, planets and stars. Amongst its many other uses it was used by Muslims to find the direction of Mecca.
Board of Longitude	The full name for the Board was Commissioners for the Discovery of the Longitude at Sea and was set up purely as a result of accidents (tragedies) due to the lack of knowledge of precise location. The dedicated aim of many was to use the position of the stars and Moon that could have taken many more decades had it not been for the invention of the marine chronometer by John Harrison.
Clepsydra	A water clock or a mechanism for drawing water from a container too large from which to pour.
Diptych	A diptych literally means two leaves and refers to a book-like object with two panels joined by a hinge. It has many uses and applications ever since ancient times but in astronomy it refers to a sundial with a string that acts as the gnomon that appears as the diptych is opened.
Ellipticity	Flattening or squeezing of a circle of sphere results in the production of an ellipse that is the shape of the orbits of astronomical bodies. It is essential to understand the parameters of ellipses to delve deeper into some astronomical mechanisms.
Escapement	The mechanism that prevents the main movement of a clock from running away and unwinding wildly. The clicking sound is the escape wheel doing its job as it regulates the movement to a particular frequency.
Foliot	The foliot was part of the verge escapement mechanism that consisted of a horizontal bar that swung to and fro to control the movement of the verge. The oscillations could be varied by moving weights on the arms of the foliot.

Appendix 2: Glossary

Fusée	As a spring winds down it loses its force. By allowing the chain attached to the spring mechanism to pass round a drum with a reducing radius the force can become even. The drum or cone is called a fusée.
Isochronism	Isochronous is simply the keeping of regular time. The great leap forward for isochronous clocks was the adaptation of the pendulum as first described by Galileo. The time of a swing of a pendulum of a specified length will be the same for different angles of swing.
Lantern clock	These weight-driven clocks were the first commonly used clocks. They were not tremendously accurate which is why they originally had only one hand, the hour hand. A bell rang every hour.
Meridian	Meridians are imaginary circles around any globe or sphere that can be anywhere on that sphere. Tying down its geographical position leads to the Greenwich meridian, 0 degrees longitude, for international navigation or, for astronomical reference, is a Great Circle passing through the celestial poles and the zenith for a particular location.
Quadrant	A quadrant is simply an instrument that is used to measure angles up to 90 degrees. Its special use in astronomy is to precisely find the coordinates of stars and was known in ancient times. The quadrant must be firmly attached to a wall or similar. A telescope is attached for accurate measurement. A hand-held quadrant was used at sea for navigation but was superseded by the more modern sextant.
Resonance	The interaction of various parameters of celestial bodies that result in stable orbits is referred to as being in resonance. The shape of the orbits themselves can interact to produce resonance, or, if resonance is not achieved something has to give and one of the bodies may be thrown out of the system altogether.
Saros cycle	A Saros cycle is the interval between times when all three bodies, the Sun, Moon and Earth are in alignment to produce an eclipse of the Moon. It is not the same as the time between the Moon being in exactly the same position in the sky.
Sidereal	Sidereal time is based on the Earth's rate of rotation measured relative to the fixed stars.
Synodic	Synodic refers to rotation round a particular body rather than referring to the stars. (sidereal). A synodic day is the time it takes a planet (such as Earth) to revolve once relative to the body it is orbiting (such as the Sun).

Appendix 3: Bibliography

A select list of sources of information, facts and references included within this work.
1. A classic and detailed history of the orrery:
   King HC, Milburn JR (1978) Geared to the Stars, The Evolution of Planetariums, Orreries and Astronomical Clocks, University of Toronto Press.
2. An essential read for the events leading up to the invention of the Harrison chronometer:
   Sobel D (2008) Longitude, The True Story of a Lone Genius Who Solved the Greatest Scientific Problem of His Time, Harper Perennial.
3. A beautiful condensate to understand Harrison and his chronometer.
   Betts J (2011) Harrison, National Maritime Museum, London.
4. A comprehensive work of the life of Dr Robert Boyle, the chemist, that contains much detail relevant to Charles Boyle, the 4th Earl of Orrery:
   Hunter M (2009) Boyle, Between God and Science, Yale University Press.
5. Not an overview of escapements but a thoroughly detailed work for an in-depth understanding and repair of escapements except Harrison's Grasshopper that is covered in other references:
   Gazeley WJ (2008) Clock and Watch Escapements, Robert Hale Ltd.
6. A comprehensive and detailed description of the life and works of Thomas Tompion including lots about George Graham and their clocks and watches. Particularly interesting are the copies of adverts for the recovery of their lost watches:
   Symonds RW (1969) Thomas Tompion: His Life and Works, Littlehampton Book Services Ltd.
7. Plenty of details of the lives of Tompion and Graham but a most informative list of their clocks and watches is tabulated here:
   Evans J (2006) Thomas Tompion at the Dial and Three Crowns, With a Concise Check List of the Clocks, Watches and Instruments from His Workshops, The Antiquarian Horological Society.
8. For details of the parliamentary facts of Charles Boyle and a good summary of his life. Only factual listing of his having the illegitimate one son (some references mention two) and two daughters by Mrs. Swordfeger:
   Cruickshanks E, Handley S, Hayton DW (2002) The House of Commons 1690–1715, 5 Volume Set (The History of Parliament), Cambridge University Press.
9. To accompany some astronomical facts referred to in Chapter 13, 'If you were there' and elsewhere:
   Buick T (2010) Rainbow Sky, Springer.
10. A comprehensive treatment of the Antikythera Mechanism:
    Marchant J (2009) Decoding the Heavens, Windmill Books.
11. An invaluable treatise for known details of the provenance of the two proto-orreries:
    Bedini SA, (1994) In Pursuit of Provenance, Hackmann WD, Turner AJ (editors) Learning, Language and Invention: Essays Presented to Francis Maddison, pp. 54–77.

12 Many topics of his life and others are bluntly revealed in the diaries of Robert Hooke that he clearly did not intend for general knowledge or publication:
Hooke R (1994) The Diaries of Robert Hooke: The Leonardo of London, 1635–1703, Nichols RS (editor), Book Guild Ltd.
13 The full works of Samuel Pepys, 11 volumes over 5,000 pages, is available but very expensive so best viewed at specialist libraries:
Pepys S (2003) The Diary Of Samuel Pepys: A New and Complete Transcription in 11 Volumes, Latham R, Matthews W (editors), The Folio Society.
For a selection of Pepys' diaries:
Pepys S (2003) The Diaries of Samuel Pepys, Penguin Classics.
Or each volume can be purchased separately:
Pepys S (2010) The Diary of Samuel Pepys: Volume I – 1660, Latham R, Matthews W (editors), Harper Collins.
14 Many museums and libraries worth the visit include;
The Clockmakers Museum; Attached to the Guildhall library, Aldermanbury, London.
National Maritime Museum incorporating the Greenwich Royal Observatory and Caird library, Greenwich, London.
Museum of The Bank of England, Threadneedle Street, London, that includes the library/archives.
The Museum of London, London Wall, City of London.
The British Museum, Great Russell Street, London.
A magnificent private collection of clocks at Belmont House near Faversham, Kent.
Armagh Planetarium located in Armagh, Northern Ireland, close to the city centre and neighboring Armagh Observatory.
The Adler Planetarium & Astronomy Museum in Chicago, Illinois, USA, was the first planetarium built in the Western Hemisphere, is the oldest in existence today and houses the first proto-orrery.
Museum of the history of science, Broad Street, Oxford, that houses the second proto-orrey.
The Science Museum, South Kensington, London, that houses the Rowley or first orrery.

Index

A
Adler Planetarium and Astronomical Museum in Chicago, 80, 94
Airy, Sir George, 58, 231
Alsatia, 204
Anchor, 26, 31, 57
Anne Victoria of Savoy, Princess, 88
Antikythera, 10
Armagh Observatory's Human Orrery, 144
Armillary, 14, 17, 19
Arnold, John, 69, 73, 74
Aryabhata, 6
Aske, Henry, 53
Astrarium clock, 31
Astrolabe, 19, 20, 31
Astronomers Royal, 76, 259
Atterbury, Francis, Bishop of Rochester, 110, 118
Aurora, 65, 257
Australopithecus, 2

B
Babylonian, 4, 12, 22
Banking, 49
Battle of the Books, 237
Belvedere, Great Palace, 83, 88
Bentley, Dr Richard, 39, 111, 112, 113, 235
Black Death, 201
Blackfriars, 204
Board of Longitude, 67, 71, 240
Boleyn, Anne, 33
Botibol, Jesse Myer, 90
Boyle's Law, 42, 107
Bradley, James, 26, 58, 77
Brazen Bull, 111
Breguet, Abraham-Louis, 73, 135
Budgell, Eustace, 78, 262

C
Caral Supe, 9
Catherine of Braganza, 216
Celsius, Anders, 64
Chemist, 275
Churchill, John, 81
Clepsydra, 22
Collegiate Church of St Mary, Youghal, 102
Comet, 8, 42, 144
Copernicus, 26
Coronal mass ejection, CME, 253
Cromwell, 105, 216, 219, 260
Cuneiform, 3, 4, 12
Cutty Sark, 217

D
Deadbeat, 42
de Dondi, Giovanni, 31
Derrynane House, 30
Desaguliers, 88, 94, 250
Dorians, 8
Downing Street, London, 115

E
Earnshaw, Thomas, 73, 74
Egyptians, 5, 12
Enlightenment, 201, 247, 263, 265, 267, 275
Equation of time, 284
Eratosthenes, 17, 66
Erzabtei St Peter, Benedictine monastery in Salzburg, 91
Exeter, 31, 116

F
Ferguson, James, 131, 137
Flamsteed, John, 41, 42, 58, 59, 85, 247, 250

Fleet, River, 200, 201, 225
Foliot, 28, 32, 33, 35, 57
Fromanteel, 35, 270

G
Gibraltar, 241, 242
Gould, Rupert Thomas, 124
Gray, Stephen, 250
Great Fire of London, 201, 203, 207, 210, 212, 219, 238
Great storm, 258
Gridiron, 58, 76
Guild, 37, 48, 259, 267

H
Habsburg, 81, 86
Hales, Stephen, 78
Halley, Edmond, 235, 251
Hampton Court, 25
Harappans, 6
Henry VIII, 25, 31, 33, 37, 59, 208
Historia Coelestis, 248
Hofburg Palace, 285
Homo sapiens sapiens, 2
Horrocks, Jeremiah, 42, 107
Hugenots, 207, 208, 269
Huygens, Christaan, 26, 34

I
Isles of Scilly, 67

J
Jones, Brothers William and Samuel, 126, 136

K
Kendall, Larcum, 69
Kentucky Vietnam Veterans Memorial, 13
Kit-Kat club, 245
Knibb, Joseph, 37
Knole House, 24, 111
Kratzer, Nicholas, 25

L
Lagadha, 6
Lantern clock, 24
Layer Christopher (Kit), 118
Leonard de Vinci, 12
Lismore Castle, 106
Longitude, 17, 56, 58, 59, 60, 66, 67, 69, 108, 124, 130, 239, 240

M
Malplaquet, battle, 83
Maria Theresia, 90, 94, 285
Marlborough, Duke of, 55, 81, 87, 106, 115
Marlowe, Christopher, 100
Martin, Benjamin, 283
Maskelyne, Nevil, 68, 126, 240
Maunder Minimum, 254, 266
Maupertuis, Pierre Louis, 64
Maya, 9
Maypole, 26
Meridian, 17, 25, 58, 76
Mesopotamia, 3
Michaelbeueren, Benedictine monastery in Salzburg, 91
Mississippian, 9
Molyneux, Samuel, 64
Moore, Sir Jonas, 42, 47, 49, 51, 57, 59, 246, 248
Moore, Sir Patrick, 136
Mudge, Thomas, 70, 71, 267
Mycenaeans, 8

N
Nesstfelt, Johann Georg, 90
Newton, Sir Isaac, 26, 39, 77, 122, 129, 212, 272

O
Orion Nebular, 9
Otford Solar System, 145
Ottery-St-Mary, 31
Oursian, Nicholas, 25, 26

P
Phalaris, Epistles of, 111, 234, 245
Pirates, 241, 263
Plane Tree, 230
Pound, Rev James, 26
Prague, 30, 32
Preston village, Kent, 99
Printing, 201, 264
Ptolemy, Claudius, 10
Punch and Judy, 238

R
Raleigh, Sir Walter, 102, 103, 106
Ramsden, Jesse, 125, 126
Richard of Wallingford, 30
Rittenhouse, David, 135
Rømer, Ole Christensen, 60

Romeyn de Hooghe, 273
Royal Mint, 268, 272

S
Sackville, Sir Richard, 97, 111
Salisbury, 31
Sexagesimal, 5
Sextant, 61, 66, 85
Shang Dynasty, 8
Sharp, Abraham, 248
Shovell, Sir Cloudesley, 67, 240, 241, 242, 259, 264, 271
Sieur de St. Pierre, 66
Slaves, 260, 263
Spanish Armada, 59
Stadtpalais, 88
Steele, Sir Richard, 87, 88, 116, 125, 243
Stowe, John, 206, 210
Strype, John, 206, 209
Sumer, 3
Sumerians, 4, 41
Swift, Jonathan, 234, 235
Swordfeger, Margaret, 115, 119, 241

T
Time ball, 231, 232, 233
Towneley, Richard, 41, 42, 57, 107
Transit, 60, 62, 63, 76
Tycho Brahe, 60, 221, 248
Tychonic system, 17

U
Urseau, 25

V
Vayringe, Philip, 94
Venus, 9, 42, 86, 131, 135, 137, 144, 240, 286
Venus Tablet of Ammisaduqua, 12
Verge, 28, 31, 56, 64
Vitruvius, 12, 22

W
Walpole, Sir Robert, 116, 118
War of the Spanish Accession, 67
Water clock, 22
Wells, 31
Wherries, 209, 279, 280
William the Conqueror, 97
World Heritage Site of Stonehenge in Wiltshire, England, 3
Wright, Thomas, 55, 86, 131, 132, 222
Wynkyn de Worde, 201

Y
Youghal, 99, 102

Z
Zapotec, 9